SpringerBriefs in Speech Technology

Studies in Speech Signal Processing, Natural Language Understanding, and Machine Learning

Series Editor:
Amy Neustein

SpringerBriefs present concise summaries of cutting-edge research and practical applications across a wide spectrum of fields. Featuring compact volumes of 50 to 125 pages, the series covers a range of content from professional to academic. Typical topics might include:

- A timely report of state-of-the-art analytical techniques
- A bridge between new research results, as published in journal articles, and a contextual literature review
- A snapshot of a hot or emerging topic
- An in-depth case study or clinical example
- A presentation of core concepts that students must understand in order to make independent contributions

Briefs are characterized by fast, global electronic dissemination, standard publishing contracts, standardized manuscript preparation and formatting guidelines, and expedited production schedules.

The goal of the **SpringerBriefs in Speech Technology** series is to serve as an important reference guide for speech developers, system designers, speech engineers and other professionals in academia, government and the private sector. To accomplish this task, the series will showcase the latest findings in speech technology, ranging from a comparative analysis of contemporary methods of speech parameterization to recent advances in commercial deployment of spoken dialog systems.

More information about this series at http://www.springer.com/series/10043

Nilanjan Dey • Amira S. Ashour
Waleed S. Mohamed • Nhu Gia Nguyen

Acoustic Sensors for Biomedical Applications

 Springer

Nilanjan Dey
Department of Information Technology
Techno India College of Technology
Kolkata, India

Waleed S. Mohamed
Department of Internal Medicine
Faculty of Medicine, Tanta University
Tanta, Egypt

Amira S. Ashour
Department of Electronics and Electrical
Communications Engineering
Faculty of Engineering, Tanta University
Tanta, Egypt

Nhu Gia Nguyen
Graduate School
Duy Tan University
Da Nang City, Vietnam

ISSN 2191-737X   ISSN 2191-7388   (electronic)
SpringerBriefs in Speech Technology
ISBN 978-3-319-92224-9   ISBN 978-3-319-92225-6   (eBook)
https://doi.org/10.1007/978-3-319-92225-6

Library of Congress Control Number: 2018943407

Printed on acid-free paper

This Springer imprint is published by the registered company Springer Nature Switzerland AG.
The registered company address is: Gewerbestrasse 11, 6330 Cham, Switzerland

Preface

Acoustic biomedical signal, electric biomedical signals, magnetic biomedical signals, mechanical biomedical signals, optical biomedical signals, and chemical biomedical signals are different types of the biomedical signals used in several healthcare monitoring applications. In these applications, it is indispensable to analyze and process the acquired biomedical signals for different diseases' prediction, detection, and monitoring. Acoustic biomedical sensors have a significant role to acquire the different acoustic biosignals from the human body [1–5]. Typically, acoustics is considered a well-developed scientific domain based on physics and expanding its scope over time into biomedical signal processing, speech, and hearing sciences. One of the vital forms of the acoustic biomedical signals is the phonocardiogram (PCG), which records the heart sound in a waveform for further visual inspection. This acoustic biomedical signal requires preprocessing stage for the signal enhancement based on automatic biomedical signal processing techniques. Furthermore, multichannel signal processing requires indispensable techniques for biomedical signal processing [6–14]. Manual investigation of these signals is time consuming and complex. Biomedical signal processing has several outstanding techniques based on either time or frequency domain that depends on the biomedical signal characteristics. Automated monitoring of the biomedical signal becomes crucial. Several applications in the acoustical realm, including the speech acoustic and the biomedical signal acoustic, are combined with signal processing and soft computing techniques to improve the biomedical signal processing.

The book supports designers, engineers, researchers, and physicians in several interdisciplinary areas to support healthcare based on the different applications of the acoustic biomedical sensors. *The book presents* an overview of the different biomedical signal types, while focusing on the acoustic biomedical signal and the body sounds. *The book emphasizes* a brief outline of the different medical sensors for biomedical signal acquisition with extensive study of the acoustic biomedical sensor applications. *This book is considered* the milestone in providing an overview on the development of the PCG biosignal and its analysis. *The book includes* several applications and real-life acoustic biomedical sensors for supporting the healthcare. Furthermore, the *book reports* the challenges facing the acoustic biomedical signal processing techniques in healthcare.

The book's inimitable features include the following:

- Offers a cutting-edge platform on the concepts and types of the biomedical signals, especially the acoustic biomedical signal
- Emphasizes the diverse approaches to analyze the heart sound signals and the different body sounds
- Provides a brief outline of the acoustic biomedical sensors as well as the other different biomedical sensors and devices
- Discovers the role of biomedical signal processing in healthcare
- Presents the heart sound detection using the acoustic biomedical sensors for further analysis to support the clinical diagnosis
- Introduces the different techniques of the biomedical signal processing for healthcare monitoring
- Discusses some applications of acoustic biomedical sensors and signal processing for prediction, detection, and monitoring of some diseases from the phonocardiogram (PCG) signal analysis
- Presents the new technologies for the design and fabrication of the acoustic biomedical sensors
- Introduces the challenges of acoustic biomedical signals as well as the gaps between the contemporary approaches of the heart sound signal analysis and their applications for clinical diagnosis

West Bengal, India   Nilanjan Dey
Tanta, Egypt   Amira S. Ashour
Tanta, Egypt   Waleed S. Mohamed
Da Nang, Vietnam   Nhu Gia Nguyen

References

1. Coté, G. L., Lec, R. M., & Pishko, M. V. (2003). Emerging biomedical sensing technologies and their applications. *IEEE Sensors Journal, 3*(3), 251–266.
2. Ma, Y., Zheng, Q., Liu, Y., Shi, B., Xue, X., Ji, W., et al. (2016). Self-powered, one-stop, and multifunctional implantable triboelectric active sensor for real-time biomedical monitoring. *Nano Letters, 16*(10), 6042–6051.
3. Dey, N., Ashour, A. S., Shi, F., Fong, S. J., & Sherratt, R. S. (2017). Developing residential wireless sensor networks for ECG healthcare monitoring. *IEEE Transactions on Consumer Electronics, 63*(4), 442–449.
4. Safari, A., & Akdogan, E. K. (Eds.). (2008). Piezoelectric and acoustic materials for transducer applications. Springer Science & Business Media
5. Chen, I. M., Phee, S. J., Luo, Z., & Lim, C. K. (2011). Personalized biomedical devices & systems for healthcare applications. *Frontiers of Mechanical Engineering, 6*(1), 3–12.
6. Tran, B. (2008). *U.S. Patent Application No. 11/480,206.*
7. Abbas, A. K., & Bassam, R. (2009). Phonocardiography signal processing. *Synthesis Lectures on Biomedical Engineering, 4*(1), 1–194.

8. Gavriely, N., & Intrator, N. (2010). *U.S. Patent No. 7,819,814*. Washington, DC: U.S. Patent and Trademark Office.
9. Barma, S., Kuan, T. W., Wu, J. S., Tseng, S. P., & Wang, J. F. (2013, March). A review on heart sound modeling: Fluid dynamics and signal processing perspective. In *Orange Technologies (ICOT), 2013 International Conference on* (pp. 201–204). IEEE.
10. Feng, D. D. (Ed.). (2011). *Biomedical information technology*. Burlington: Academic Press.
11. Zouridakis, G. (2003). *Biomedical technology and devices handbook*. Boca Raton: CRC Press.
12. Mendes, J. J. A., Jr., Vieira, M. E. M., Pires, M. B., & Stevan, S. L., Jr. (2016). Sensor fusion and smart sensor in sports and biomedical applications. *Sensors, 16*(10), 1569.
13. KO, W. (1988). Biomedical transducers. In J. Kline (Ed.), *Handbook of biomedical engineering* (pp. 3–71). New York: Academic Press.
14. Ahmed, S. S., Dey, N., Ashour, A. S., Sifaki-Pistolla, D., Bălas-Timar, D., Balas, V. E., & Tavares, J. M. R. (2017). Effect of fuzzy partitioning in Crohn's disease classification: A neuro-fuzzy-based approach. *Medical & Biological Engineering & Computing, 55*(1), 101–115.

Abstract

The interdisciplinary acoustic biosignals nature and the biomedical sensors are challenging. In this book, application-related studies to the acoustic biomedical sensors are covered in depth. This book attracts engineers, designers, researchers, and physicians in innumerable interdisciplinary areas, including acoustic biomedical signal analysis, engineering, healthcare, biomedical signal processing, and patient monitoring. The book hosts the concept of a wide spectrum of different biomedical signals, including acoustic biomedical signals as well as the thermal biomedical signals, magnetic biomedical signals, and optical biomedical signals to support healthcare. The signal processing approaches, such as *filtering*, *Fourier transform*, *spectral estimation*, and *wavelet transform*, to the biomedical signals are also explored and discussed, for example, the analysis of cardiac signals, breathing cycle, and the heart sound. The book provides global outstanding research and recent progress in some applications of acoustic biomedical sensors and biosignal processing for prediction, detection, and monitoring of some diseases from the phonocardiogram (PCG) signal analysis. Several challenges and future perspectives related to the acoustic sensor applications are highlighted in this book.

Acknowledgment

We are thankful to our *parents* and *families* for their support throughout the project.

A special thanks to the Springer-publisher team and Prof. Amy Neustein, the series editor, for her immense support from the very beginning.

Last but not the least, we would like to thank our readers, anticipating that they will find the book as a valuable resource in their domain.

Author(s)

Contents

About the Authors

Nilanjan Dey received his B.Tech. degree in Information Technology from West Bengal University of Technology in 2005, M.Tech. in Information Technology in 2011 from the same university, and Ph.D. in Digital Image Processing in 2015 from Jadavpur University, India. In 2011, he was appointed as an Assistant Professor in the Department of Information Technology at JIS College of Engineering, Kalyani, India, and then at Bengal College of Engineering and Technology, Durgapur, India, in 2014. He is now employed as an Assistant Professor in the Department of Information Technology, Techno India College of Technology, India. His research topic is about signal processing, machine learning, and information security.

Dr. Dey has 25 books and about 350 published journal papers. He is an Associate Editor of IEEE ACCESS and is currently the Editor-in-Chief of the *International Journal of Ambient Computing and Intelligence* and *International Journal of Rough Sets and Data Analysis*, Co-Editor-in-Chief of the *International Journal of Synthetic Emotions* and *International Journal of Natural Computing Research*, and Series Editor of *Advances in Geospatial Technologies* Book Series and Co-Editor of *Advances in Ubiquitous Sensing Applications for Healthcare (AUSAH)* Elsevier (Book Series).

Amira S. Ashour received her B.Eng. degree in Electrical Engineering in 1997, M.Sc. in Image Processing in 2001, and Ph.D. in Smart Antenna in 2005 from the Faculty of Engineering, Tanta University, Egypt. In 2005, she was appointed as a Lecturer in Electronics and Electrical Communications Engineering at Tanta University. Dr. Amira is currently an Assistant Professor and the Head of the Department of Electronics and Electrical Communications Engineering, Faulty of Engineering, Tanta University, Egypt. She has been the Vice Chair of Computer Engineering Department, Computers and Information Technology College, Taif University, KSA, for 1 year from 2015. She has also been the Vice Chair of

CS Department, CIT College, Taif University, KSA, for 5 years. Her research interests are smart antenna, direction of arrival estimation, targets tracking, image processing, medical imaging, machine learning, biomedical systems, pattern recognition, signal/image/video processing, image analysis, computer vision, and optimization. She has 8 books and about 100 published journal papers. She is an Editor-in-Chief of the *International Journal of Synthetic Emotions (IJSE)*, IGI Global, USA. She is an Associate Editor of the *IJRSDA* as well as the *IJACI*, IGI Global, USA. She is an Editorial Board Member of the *International Journal of Image Mining (IJIM)*, Inderscience.

Waleed S. Mohamed   received his M.Sc. in Internal Medicine in 1993 and his M.D. in Internal Medicine in 2000 from the Faculty of Medicine, Tanta University, Egypt. He is also a Professor of Internal Medicine at the Faculty of Medicine, Tanta University, Egypt, from 2011 till now. Prof. Waleed is currently the Head of the Gastroenterology and Hepatology Unit, in the mentioned faculty. He is a Consultant of Gastroenterology and Diabetology and of Ultrasonography and Color Doppler at Tanta University, Faculty of Medicine, Tanta University, Egypt, and previously the same in Taif University, KSA, for 10 years from 2006 till 2016. He was a Professor and the Chairperson of the Human Resource Unit, Taif College of Medicine, Taif University, KSA, for 10 years. He was the Medical Director at Taif University Medical Outpatient Clinics, KSA, for 10 years since 2006. He is a member of the editorial board of the *Immunoendocrinology* and *Journal of Liver and Pancreatic Diseases (JLPD)*. He is a reviewer in several international journals. His research interests include hepatology, gastroenterology, nonalcoholic fatty liver disease, obesity, diabetes mellitus, herbal medicine, medical education, and healthcare quality. He has 2 books and 50 published papers in international journals and conferences. He is a life member of the Egyptian Society of Blood Diseases; Arab Society for Studying Liver Diseases; Egyptian Association of Kidney; Egyptian Association of Liver, Gastrointestinal, and Tropical Diseases; and Egyptian Association of Diabetes and Metabolism.

Nhu Gia Nguyen   received his Ph.D. degree in Computer Science from Hanoi University of Science, Vietnam National University, Vietnam. Now, he is the Vice Dean of the Graduate School at Duy Tan University, Vietnam. He has a total academic teaching experience of 18 years with more than 50 publications in reputed international conferences, journals, and online book chapter contributions (indexed by: SCI, SCIE, SSCI, Scopus, ACM, DBLP). His area of research includes network communication, security and vulnerability, network performance analysis and simulation, cloud computing, and biomedical image processing. Recently, he has been part of the technique program committee and the technique reviewer and the track chair for international conferences: FICTA 2014, ICICT 2015, INDIA 2015, IC3T 2015, INDIA 2016, FICTA 2016, IC3T 2016, IUKM 2016, and INDIA 2017, under Springer-ASIC/LNAI Series. Presently he is an Associate Editor of the *International Journal of Synthetic Emotions (IJSE)*.

Chapter 1
Introduction

The interface between the physical sciences, electronics, and life sciences becomes inhabited by several researchers to fulfill the needs of the medical/life scientist in the biomedical community. Based on the chemical, biological, and physical principles, the instruments improvement burgeoned. Conversely, the analytical instruments require several types of sensors extending from elementary devices for temperature and flow measurements, nonionizing and ionizing radiation to biological, chemical, ultrasound, and acoustic sensing transducers [1–9].

Monitoring real-time biomedical signals is the clue for superior management, earlier detection, prediction, and diagnosis of chronic diseases, including strokes and heart attacks. Different biomedical signals can be acquired from different medical sensors, which require real-time biomedical signal processing and analysis for improving healthcare and managing critical care situations. Different types of biomedical signals can be acquired to reflect the patient's status. These biomedical signals include temperature records, voltage record by an electrode placed on the scalp, phonocardiogram (PCG) acoustic, electrocardiogram (ECG) showing the heart's electrical activity, and the electroencephalogram (EEG) signals, which show the electrical activity of the brain. Several obtained parameters are unreliable, which become challenging for the electronics' designers to process and deploy these signals. However, there are common characteristics of the biomedical signals even with their variation in terms of the environmental conditions, including the design of the equipment, electrode positioning, and the existence of the fats and/or blood vessels under which they are obtained. Advancement in technology, electronic device design/fabrication, and sensors provided more biomedical signals for monitoring, predicting, and detecting different diseases, according to the acquired biomedical signal characteristics, from the specific associated sensors, for further signal processing, which is the energetic phase. Multichannels of the biomedical signals are produced from the different sensors that are connected to the patient for acquiring specific information about certain diseases [10–18].

N. Dey et al., *Acoustic Sensors for Biomedical Applications*,
SpringerBriefs in Electrical and Computer Engineering,
https://doi.org/10.1007/978-3-319-92225-6_1

In medical, industrial, and engineering applications, there is an emerging need for reliable, small, inexpensive, and disposable sensors. Sensors are one of the wildest rising markets. Especially, the biosensor market is progressively promising due to their application in biotechnology, healthcare, and medical applications, such as in the early cancer detection, glucose testing, and physiologically/pathogen detection as a prevailing tool for early disease diagnosis and treatment.

A sensor can be defined as a device that reacts to an input under concern as it measures or detects certain property, records, or condition according to the received information [19]. Generally, the device which only detects some property or condition based on the absence or presence of a physical amount is a detector not a sensor. Detectors have a vital role in medicine, particularly alarms. In biomedical applications, a sensor is known as a responding device to a physical input of interest recording the related optical/electrical output. The physical and biomedical input includes biochemical concentrations/quantities and any medical vital signs from the patients. The sensors' system is a device with an electrical output that indicates/records or indicates the electrical signals, which are amplified and processed for further final output to a chart recorder, displaying monitor or input to a storage system. However, numerous widely used sensors do not display a linear performance. Both transducer and sensor are synonymous terms; nevertheless, the transducer is known as a transformation device of the energy from one form to another. Sometimes, the transducers can be considered components of the sensors, such as the diaphragm in a microphone that converts the sound energy into strain energy, and then a second transducer is essential to convert this energy to a recordable electrical energy to produce the whole sound sensor [20–24].

Biosensors are analytical devices that use a biological identification system to target macromolecules or molecules. They include a physiochemical sensor (transducer) that transforms the biological signal from the bio-recognition system to an assessable and measured signal. A biosensor consists of three modules, namely, the detector to identify the stimulus, the transducer to convert the stimulus to an output, and the output system to amplify and display the output in a proper format. The piezoelectricity is a phenomenon that appears in specific crystals, such as Rochelle salt and quartz, where the mechanical stress convinces voltage generation and vice versa [25, 26].

One of the most imperative sensors is the acoustic sensor, which detects/collects the acoustic biosignals (bioacoustic). These acoustic biosignals ascend from the body's vital functions conveying physiological data that indicates the state of health and the cardiorespiratory pathologies. Several acoustic biosignals are initiated inside the human body, such as the snoring sounds, lung sounds, and heart sounds, where any vibrating structure in the body produces acoustic sounds. These acoustic sounds are inhibited during the propagation via the tissues of the skin. The dissimilar sounds of the body interfere together inducing mechanical skin vibrations that are observed by the sound sensor for further conversion into the electrical signal. From clinical and engineering point of view, it is highly informative to study the origin of the sounds. Extensive studies related to sound physics as well as biology have been conducted.

In order to observe and record the acoustic biosignals, biosensors, which are analytic transducer devices, can convert the biomedical information into a detectable signal. These acoustic biosensors consist of (i) a detector for identifying the stimulus, (ii) transducer for converting the stimulus to an output, and (iii) the output system for amplifying and displaying the output. The acquired biosignals by the acoustic biomedical sensors are then processed using biomedical signal-processing techniques for extracting the valuable information from the measured data (biomedical signals) after de-noising and enhancement. In addition, biomedical signal conditioning, pattern recognition, classification, and biomedical signal compression can be performed for accurate diagnosis [27–34].

An acoustic wave biosensor employs mechanical or acoustic waves as a recognition mechanism to attain biophysical, biochemical, and medical information. It senses changes in conductivity, elasticity, mass, and dielectric properties of the electrical or mechanical variations. Such biosensors employ the piezoelectric effect to stimulate electrically the acoustic waves at the input transducer and then to receive the generated waves at the output transducer. In the piezoelectricity effect, a voltage is produced on the surface of the piezoelectric material due to the compression of several piezoelectric crystals. Acoustic biosensors can be prepared using piezoelectric crystals, such as lithium tantalite, lithium niobate, or quartz, as they are environmentally stable and robust. Furthermore, such sensors can detect several versatile biomolecules. Acoustic wave sensors can be classified based on the generated waves into surface or bulk acoustic waves. Each of them has advantages/disadvantages based on applications under consideration [35–38]. Acoustic wave devices have marvelous features to be employed in sensor systems for medical diagnosis. Such acoustic wave sensors are configurable to many applications, sensitive, easily portable, and utilized as actuators.

The organization of the remaining chapters is as follows. Chapter 2 contains a broad description of the biomedical signals, including their characteristics and classification, as well as the definition of the biosensors and the biomedical signal acquisition process is also mentioned. Chapter 3 introduces the concept of the acoustic wave technology and the fundamentals of acoustics and psychoacoustics. In addition, the different sources of the acoustic biosignals in the human body are reported, including the heart sounds, breath sound, gurgling/intestinal sound, Korotkoff sounds, vascular sounds, and the friction rub sounds. Chapter 4 provides a brief highlight on the different acoustic sensors and the related concepts and requirements. In this chapter the piezoelectricity effect, the acoustic sensor design, and the acoustic stethoscope sensor are introduced. The principle of the acoustic wave sensors, including the difference between the bulk acoustic wave sensors and the surface acoustic wave sensors, is presented. Chapter 5 discussed in brief the acoustic sensors for biomedical applications, such as the acoustic waveguide sensor for chemical detection and the stethoscopic sensor for respiratory sound recording. Furthermore, the role of the signal analysis is highlighted briefly under a heart sound analysis in clinical diagnosis discussion. Finally, the book concludes in Chap. 6.

References

1. Campajola, L., & Di Capua, F. (2016). Applications of accelerators and radiation sources in the field of space research and industry. *Topics in Current Chemistry, 374*(6), 84.
2. Duan, X., Huang, Y., Cui, Y., Wang, J., & Lieber, C. M. (2001). Indium phosphide nanowires as building blocks for nanoscale electronic and optoelectronic devices. *Nature, 409*(6816), 66.
3. Hughes, P. G., Votava, O., West, M. B., Zhang, F., & Kable, S. H. (2005). Pulsed oscillating mass spectrometer: A miniaturized type of time-of-flight mass spectrometer. *Analytical Chemistry, 77*(14), 4448–4452.
4. Rivetti, A. (2015). *CMOS: Front-end electronics for radiation sensors* (Vol. 42). Boca Raton: CRC Press.
5. Chatterjee, S., Hore, S., Dey, N., Chakraborty, S., & Ashour, A. S. (2017). Dengue fever classification using gene expression data: A PSO based artificial neural network approach. In *Proceedings of the 5th international conference on frontiers in intelligent computing: Theory and applications* (pp. 331–341). Singapore: Springer.
6. Dey, N., Ashour, A. S., Shi, F., & Sherratt, R. S. (2017). Wireless capsule gastrointestinal endoscopy: Direction-of-arrival estimation based localization survey. *IEEE Reviews in Biomedical Engineering, 10*, 2–11.
7. Ashour, A. S., & Dey, N. (2016). Adaptive window bandwidth selection for direction of arrival estimation of uniform velocity moving targets based relative intersection confidence interval technique. *Ain Shams Engineering Journal.*
8. Skoog, D. A., Holler, F. J., & Crouch, S. R. (2017). *Principles of instrumental analysis.* New York: Cengage Learning.
9. Franssila, S. (2010). *Introduction to microfabrication.* Chichester: Wiley.
10. OKOYE, G. C. (2008). Biomedical technology and health human life. *Biomedical Engineering, 1*, 12.
11. Castano, L. M., & Flatau, A. B. (2014). Smart fabric sensors and e-textile technologies: A review. *Smart Materials and Structures, 23*(5), 053001.
12. Sun, Y., & Yu, X. B. (2016). Capacitive biopotential measurement for electrophysiological signal acquisition: A review. *IEEE Sensors Journal, 16*(9), 2832–2853.
13. Korotcenkov, G. (Ed.). (2011). *Chemical sensors: Comprehensive sensor technologies volume 6: Chemical sensors applications* (Vol. 6). New York: Momentum Press.
14. Gospodinova, E., Gospodinov, M., Dey, N., Domuschiev, I., Ashour, A. S., & Sifaki-Pistolla, D. (2015). Analysis of heart rate variability by applying nonlinear methods with different approaches for graphical representation of results. *Analysis, 6*(8).
15. RajaRajeswari, P., Raju, S. V., Ashour, A. S., Dey, N., & Balas, V. E. (2016, June). Active site cavities identification of amyloid beta precursor protein: Alzheimer disease study. In *Intelligent Engineering Systems (INES), 2016 IEEE 20th Jubilee International Conference on* (pp. 319–324). IEEE.
16. Kamal, M. S., Chowdhury, L., Khan, M. I., Ashour, A. S., Tavares, J. M. R., & Dey, N. (2017). Hidden Markov model and Chapman Kolmogrov for protein structures prediction from images. *Computational Biology and Chemistry, 68*, 231–244.
17. Liu, K. K., Wu, R. G., Chuang, Y. J., Khoo, H. S., Huang, S. H., & Tseng, F. G. (2010). Microfluidic systems for biosensing. *Sensors, 10*(7), 6623–6661.
18. Nichols, S. P., Koh, A., Storm, W. L., Shin, J. H., & Schoenfisch, M. H. (2013). Biocompatible materials for continuous glucose monitoring devices. *Chemical Reviews, 113*(4), 2528–2549.
19. Eggins, B. R. (2008). *Chemical sensors and biosensors* (Vol. 28). Chichester: Wiley.
20. Graf, R. F. (1999). *Modern dictionary of electronics.* Oxford: Newnes.
21. De Marcellis, A., & Ferri, G. (2011). *Analog circuits and systems for voltage-mode and current-mode sensor interfacing applications.* Springer Science & Business Media.
22. Karaa, W. B. A., Mannai, M., Dey, N., Ashour, A. S., & Olariu, I. (2016, August). Gene-disease-food relation extraction from biomedical database. In *International workshop soft computing applications* (pp. 394–407). Cham: Springer.

23. Chatterjee, S., Dey, N., Shi, F., Ashour, A. S., Fong, S. J., & Sen, S. (2017). Clinical application of modified bag-of-features coupled with hybrid neural-based classifier in dengue fever classification using gene expression data. *Medical & Biological Engineering & Computing*, 1–12.
24. Holford, S. K. (1981). Discontinuous adventitious lung sounds: measurement, classification, and modeling.
25. Collins, S. A. (1990). *Sensors for structural control applications using piezoelectric polymer film* (Doctoral dissertation, Massachusetts Institute of Technology).
26. Hebra, A. J. (2010). Acoustics. In *The physics of metrology* (pp. 271–299). Vienna: Springer.
27. Dufresne, J. R., Carim, H. M., & Drummond, T. E. (2013). *U.S. Patent No. 8,548,174*. Washington, DC: U.S. Patent and Trademark Office.
28. Semmlow, J. L., & Griffel, B. (2014). *Biosignal and medical image processing*. Boca Raton: CRC press.
29. Patel, H. K. (2016). *The electronic nose: Artificial olfaction technology*. Ahmedabad: Springer.
30. Dey, N., & Ashour, A. S. (2018). Microphone array principles. In *Direction of arrival estimation and localization of multi-speech sources* (pp. 5–22). Cham: Springer.
31. Dey, N., & Ashour, A. S. (2018). Challenges and future perspectives in speech-sources direction of arrival estimation and localization. In *Direction of arrival estimation and localization of multi-speech sources* (pp. 49–52). Cham: Springer.
32. Dey, N., Ashour, A. S., Shi, F., Fong, S. J., & Tavares, J. M. R. (2018). Medical cyber-physical systems: A survey. *Journal of Medical Systems, 42*(4), 74.
33. Dey, N., & Ashour, A. S. (2017). Ambient intelligence in healthcare: A state-of-the-art. *Global Journal of Computer Science and Technology*.
34. Kumar, L. A., & Vigneswaran, C. (2015). *Electronics in textiles and clothing: Design, products and applications*. Boca Raton: CRC Press.
35. Nihonyanagi, S., Eftekhari-Bafrooei, A., Hines, J., & Borguet, E. (2008). Self-assembled monolayer compatible with metal surface acoustic wave devices on lithium niobate. *Langmuir, 24*(9), 5161–5165.
36. Fu, Y. Q., Luo, J., Flewitt, A., Walton, A., Desmulliez, M., & Milne, W. (2016). Piezoelectric zinc oxide and aluminum nitride films for microfluidic and biosensing applications. *Biological and Biomedical Coatings Handbook Applications*, 335.
37. Caliendo, C., Contini, G., Fratoddi, I., Irrera, S., Pertici, P., Scavia, G., & Russo, M. V. (2007). Nanostructured organometallic polymer and palladium/polymer hybrid: surface investigation and sensitivity to relative humidity and hydrogen in surface acoustic wave sensors. *Nanotechnology, 18*(12), 125504.
38. Campifelli, A., Bartic, C., Friedt, J. M., De Keersmaecker, K., Laureyn, W., Francis, L., Frederix, F., Reekmans, G., Angelova, A., Suls, J., Bonroy, K., De Palma, R., Cheng, Z., & Borghs G. (2003, September). Development of microelectronic based biosensors. In *Custom Integrated Circuits Conference, 2003. Proceedings of the IEEE 2003* (pp. 505–512). IEEE.

Chapter 2
Biomedical Signals

Abstract In our daily life, sensors are corporate in several devices and applications for a better life. Such sensors as the tactile sensors are included in the touch screens and the computers' touch pads. The input of these sensors is from the environment that converted into an electrical signal for further processing in the sensor system. The sensor's main role is to measure a specific quantity and create a signal for interpretation. The human bodies continuously communicate health information that reflects the status of the body organs and the overall health information. Such information is typically captured by physical devices that measure different types of information, such as measuring the brain activity, blood glucose, blood pressure, heart rate, nerve conduction, and so forth. According to these measurements, physicians decide the diagnosis and treatment decisions. Engineers are realizing new acquiring devices to measure noninvasively the different types of signals for further analysis using mathematical algorithms and formulae. This chapter includes classifications of the biosignals based on several principles. In addition, the different biosensors are highlighted including the role of the biopotential amplifier stage within the sensor system. Finally, the biomedical signal acquisition and processing phases are also included.

2.1 Classifications and Characteristics of Biomedical Signals

All biomedical systems produce signals to influence the human body or analyze biosignal to extract useful information about the functioning of the human body. The signal can be defined generally as the observed parameter from any object; specifically the biomedical signal represents the physiological phenomenon description of any living objects. Biomedical signal/biosignal is the signal that conveys biological information about the state or behavior of the living objects. The extracted information in the biomedical signal can be simple, such as the human blood pressure and the wrist pulse, or complex, such as the information obtained from the

© The Author(s), under exclusive licence to Springer International Publishing
AG, part of Springer Nature 2019
N. Dey et al., *Acoustic Sensors for Biomedical Applications*,
SpringerBriefs in Electrical and Computer Engineering,
https://doi.org/10.1007/978-3-319-92225-6_2

analysis of the internal soft tissues' structure using ultrasound scanner. Biosignals are used to realize the underlying physiological mechanisms of certain biological system or event.

The main biomedical signal types include the signal, carotid pulse (CP) signal, electrocardiogram (ECG) signal, electroencephalogram (EEG) signal, phonocardiogram (PCG) signal, and speech signals. Mostly, there are five sources of noise that affect the biosignals, including the aliasing, interference, thermal noise, sampling noise, instrument noise, and power line alternative current (AC) [1–10]. Several classifications are raised to categorize the biosignals, according to biosignal source, number of channels, dimensionality, biosignal model, and nature as illustrated in Fig. 2.1.

Figure 2.1 illustrates the different classifications of the biosignals, according to certain criteria, which is explained as follows [11–20]:

(i) Based on the system of the biosignal origin, where the biosignals vary from any other signals in terms of the application as the biosignals originate from different sources. Thus, the biosignals can be categorized into the signals from auditory system, nervous system, cardiovascular system, endocrine system, circulatory system, musculoskeletal system, vision system, gastrointestinal system, and respiratory system.

(ii) Based on the required number of channels to acquire a specific biosignal, which can be one channel to display the pulse wave, three channels to display the accelerometer data, or multichannel with the electroencephalography (EEG) signals, for example.

(iii) Based on the dimensionality, the biosignals can be 1D (one-dimensional), such as the phonocardiogram (PCG) signal, the electrocardiogram (ECG) signal, and the electroencephalogram (EEG) signal; 2D (two-dimensional), such as the temperature map; 3D (three-dimensional), such as the magnetic resonance (MR) images; or 4D (four-dimensional), such as the functional magnetic resonance (fMR) images.

(iv) Based on the signal models (analysis approach), the biosignal can be deterministic or stochastic. The deterministic biosignal is either periodic (sinusoidal or complex) or nonperiodic (transient), which is predictable. The stochastic biosignal is nondeterministic, where its state is determined by a random element and predictable actions. It can be stationary or nonstationary.

(v) Based on the physical nature of the signal, the biosignals can be categorized into chemical, optical, magnetic, electric, thermal, or acoustic (mechanical).

1. Chemical biosignals provide information about various chemical agents' concentration in the human body, such as the blood oxygen level, glucose level, breathing airflow, and gases in the blood.

2. Optical biosignals use optical approaches to sense the biochemical analyses. The biooptical signals reflect the biologic system's optical functions stirring naturally or induced by the certain measurement. Recently, the progress of the fiber-optic technology opened massive applications of the biooptical signals.

Fig. 2.1 Biosignal
classification

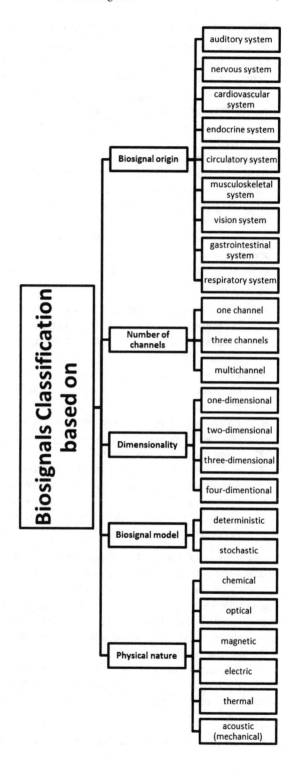

3. Magnetic biosignals are produced due to the weak magnetic fields produced by different cells and organs, such as the lungs, heart, and brain. The measurements of these magnetic fields offer significant information. Examples of such biomagnetic signals include the magnetoencephalogram (MEG) signal and the magnetoneurogram (MNG) signal produced from the neural cells as well as the magnetomyogram (MMG) signal and the magnetocardiogram (MCG) signal, e.g., produced from the muscle cells.

4. The electric signals result from the electric field due to the intra- and extracellular ionic currents produced in the organs or cells (muscle/nerve). This electric biosignal which is called also bioelectric signals or biopotentials is generated from the electrochemical mechanism in the single ionic channels that generates an action potential. Such biosignals result from the neural cells, such as the electroretinogram (ERG) signal, the electroneurogram (ENG) signal, and the electroencephalogram (EEG) signal, or from the muscle cells, such as the electromyogram (EMG) and the electrocardiogram (ECG) signal, or from other cells, such as the galvanic skin response (GSR) signal and the electrooculogram (EOG) signal [21, 22].

   4.1. ECG signal is a graphical presentation of the heart electrical activity of overtime, which is recorded by an electrocardiograph. It shows the voltage difference between the attached electrode pairs and the heart muscle activity.

   4.2. EEG signal reflects the brain electrical activity recorded from the placed electrodes on the scalp to sense pathologies related to stimulus-directed performance.

   4.3. EMG signal represents the muscles' physiologic properties of contraction and rest by detecting the electrical potential produced by the muscle cells using the electromyography.

   4.4. EOG signal is recorded using the electrooculography to measure the retina resting potential. It records the dissimilarity in the electrical charge between the back and front of the eye associated with the movement of the eyeball. The EOG signal is acquired by the electrodes positioned on the skin near the eye.

Simple transducers are used to acquire the bioelectric signals. Microelectrodes are used as sensors for single-cell measurements, where the measured action potential represents the biosignal. However, surface electrodes are involved as sensors to measure the electric field generated by the many cells' action for more unrefined measurements. Generally, the obtained bioelectric signals from the electrodes can be considered an interface between the body and the measurement instrument. Due to the specific bioelectric signals' characteristics and the device-related interferences, designing readout circuits for bioelectric signal measurements in real-time monitoring inspires several engineers to cope with various problems and to design wearable bioelectric acquisition systems.

5. The bioimpedance (IMP) signals are also produced due to the tissue's impedance that conveys information concerning the automatic nervous system activity,

endocrine activity, blood distribution, and the blood volume. The IMP measurement assists the body's properties assessment by measuring the body tissue's reaction to an induced tension. Four electrodes are generally used to measure the IMP measurements. Specifically, the position of the measuring electrode (probe) on the body surface is challenging [23–27].

6. Thermal biosignals, including the body temperature and temperature maps based on the heat absorption and heat loss in the body, or the distribution of the temperature on the body surface [28].

7. Mechanical biosignals replicate the body parts' mechanical functions, such as the chest movements during respiration process, the blood pressure, the accelerometer signals, and the phonocardiogram (PCG) signal, which reflects the heartbeats' sounds. Biomechanical signals originated from mechanical functions of the biologic system, including flow signals, tension/pressure signals, and displacement/motion signals. The biomechanical signals' measurement involves different sensors/transducers, where the mechanical phenomenon does not propagate the acoustic, magnetic, and electric fields [29–31].

8. Acoustic biosignals are defined as the biosignals produced from the respiratory sounds, cough sounds, squawk, wheeze, and snoring sounds [32, 33]. Typically, several biomedical phenomena generate acoustic noise, which offers significant information about this phenomenon. For example, the air flow in the lungs produces acoustic sounds, muscle contraction produces an acoustic noise, the blood flow in the heart produces acoustic noise, and the sounds generated in the digestive tract. These produced sounds are extensively used in medicine. Generally, the acoustic biosignals describe the acoustic sound produced by the body (motion/vibration), which is considered a subset of the mechanical biosignals. Bioacoustic signals access different body sounds, including swallowing and snoring, respiratory sounds, the cardiac sounds (phonocardiography), and the crackles of the muscles/joints. This acoustic energy transmits through the biologic medium; thus, acoustic transducers, such as the accelerometers and microphones, are used to acquire such bioacoustics signal.

Another classification of the biosignals is illustrated in Fig. 2.2.

The continuous physiological signal acquisition allows the detection and prevention of the different diseases, such as the neurological pathologies or the cardiovascular diseases. In order to acquire any of the preceding biomedical signals, different types of sensors/transducers are used according to the biomedical application as illustrated in Fig. 2.3. Biosignal acquisition is the main role of any biomedical instrument toward the most sympathetic of the human physiology using hardware, portable, and wireless acquisition transducers/sensors. However, the biosignal acquisition is insufficient, where further biomedical signal processing is compulsory to process the attained signal in order to acquire the significant information from the noisy biosignal [34–40].

The biomedical signal processing poses some complications due to the underlying system's complexity and the required noninvasive, indirect measurements. Several biosignal processing techniques are developed based on the underlying sig-

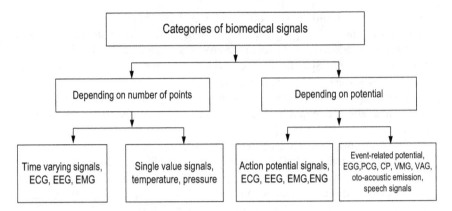

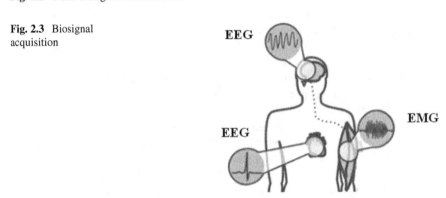

Fig. 2.2 Other biosignal classifications

Fig. 2.3 Biosignal
acquisition

nal's characteristics, the aim of the biomedical signal processing, and the test conditions. Biosignal processing consists mainly of the feature extraction phase step to extract the points of interest in the biosignal that indicates the body conditions, such as the onset of the EMG signal and the heart rate variability from the ECG signal. Over the years, several biosignals that reflect the human body conditions have been characterized and studied to transform the ways for various disease diagnoses. All the preceding biosignals are acquired using transducers/sensors of different types according to the required detected biosignal.

2.2   Biomedical Sensors

Sensors can be classified based on their sensing principle into biosensors, mechanical sensors, magnetic sensors, optical sensors, and thermal sensors. In all measurement systems and medical devices, sensors (transducers) are considered the critical components that provide a practical, electrical output in response to a definite measurand. This output reflects information about the human body. Biomedical sensors

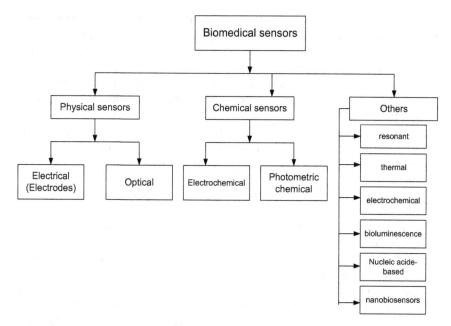

Fig. 2.4 Biomedical sensor categories

are categorized into biosensor, chemical sensor, and physical sensor as illustrated in Fig. 2.4. Biosensors are considered a type of the chemical sensors to sense antibody, antigen, hormone, enzyme, and microbes and detect biological signals [41–47]. Chemical sensors are used to detect the concentration and ingredient of liquids in the body. Physical sensors are used to measure thermal, geometric, hydraulic, and mechanical variables, such as the body temperature, blood pressure, blood flow, blood viscosity, blood flux, bone growth velocity, and muscle displacement.

Generally, the biosensors can be well-defined as analytical devices which use a biological identification system to target macromolecules or molecules. In order to convert the obtained signals from the recognition system to detectable signals, a physiochemical transducer should be included in the biosensors. The main components of the biosensors include the detector to identify the stimulus, the transducer to convert the stimulus to an output, and the output system including the amplification and display processes of the output.

There are several common characteristics that should be fulfilled in the biomedical sensors, where they should be soft, reliable, and safe as they touch the patient's inner organs or the skin. Additionally, implantable biosensor must be compatible with the human body and has a long operational lifetime. Consequently, evaluating the biomedical sensor performance and their stringent design specifications has a great impact on the accurate medical diagnosis. Several metrics are measured in order to describe the characteristics of the sensor for further selection according to the application. Such metrics include (i) the sensitivity, which is the relationship between the output change for a given input change; (ii) accuracy, which is the difference between the true and actual measured values by the sensor; (iii) the measurement range, which refers to the expected maximum/minimum operation limits of the sensor; (iv) the linearity, which refers to the maximum deviance between the calibration and the fitting curves of the sensor's measurements; (v) the hysteresis, which is the measurement's delay due to the variation direction of the measured signal; (vi) the sensor's frequency response, which represents the sensitivity variation with frequency; (vii) the signal-to-noise ratio (SNR), which represents the measured signal power ratio to the power of the noise; (viii) the drift, which is the change in the reading of the sensor with constant input; (ix) the response time, which refers to the time reserved by the sensor to reach a percent of its steady state, when its input change; (x) the resolution, which is the smallest detected discernible input change; and (xi) the offset, which refers to the output value with zero input value [48–52].

Specifically, biosensor can be classified based on the biological sensing component into immune sensors, tissue sensors, cell sensors, microbe sensors, and enzyme sensors. Another classification can be considered in terms of the biosensors' used signal converter, namely, optical biosensors, bioelectrode sensors, thermal biosensors, piezoelectric biosensor, and semiconductor biosensor. In terms of the interaction between the measured material and the sensing component, the biomedical sensors can be classified into catalytic biosensors or affinity biosensors. The foremost applications of the biomedical sensors are (i) detecting control parameters, such as measuring the concentration of enzymes to control the food fabrication; (ii) detecting clinical information, such as the blood pressure and the body temperature; and (iii) monitoring biological parameters inside/outside the body, such as the heart sound, the brain activity, and the heart activity.

Biomedical signals are recorded as voltages, electrical field strengths, and potentials that are generated by muscles/nerves. Thus, the sensors have a vital role in the various biomedical instruments to convert the acquired body signal into an electrical signal, where the electrical components and electrical circuits are used to detect the biosignal by different sensors. The bioinstrumentation is then formed after connecting the electrical components and the biosignal sensors. However, the measured biomedical signal has very low voltage levels ranging from 1 μV to 100 mV, for example, the magnitude of the EEG is in microvolt and that of the ECG is in millivolts. In addition, a high level noise and high source impedance exist and interpret the biosignal. Thus, the biosignals require amplification to be compatible with the medical instruments/devices, including the A/D (analog to digital) converters, the recorders, and the displays. In order to perform such task, amplifiers, known as biopotential amplifiers, are included to measure the biosignals and offer selective amplification to the physiological signal, interference signals, and, the reject superimposed noise [53–58]. Figure 2.5 illustrates the position of the biopotential amplifier in the medical instrument.

This assures the biosignal protection from compensations through current and voltage flow for both the electronic equipment and the patient. The accurate design of the biopotential amplifier rejects the signal interferences' large portion, espe-

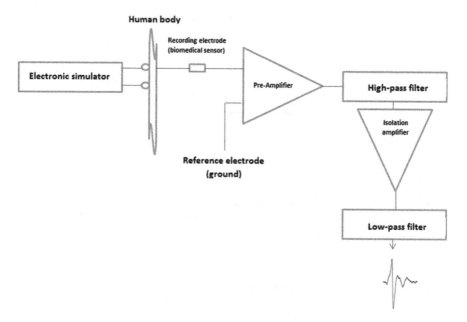

Fig. 2.5 Biopotential amplifier stages

cially the line-frequency interference with strong common-mode signal rejection. In addition, the biopotential amplifier has to offer suitable gain range and signal-to-noise ratio.

Generally, the operational amplifier is an electronic device including multi-resistors, capacitors, and transistors. It is the foundation of any bioinstrumentation as it has a significant role to amplify the weak biosignal and adjust the current or voltage in the detecting circuit. In order to retain the current down from the measured, the used amplifiers should have a very high input impedance. The preamplifier sets the phase for the biosignal quality as it eliminates/minimizes the interfering signals during the measurement of the biopotentials as shown in Fig. 2.5. After preamplifying the biosignals, several filters are included, such as the high-pass filter (HPF) and low-pass filter (LPF) to eliminate the useless signal and highlight the beneficial biosignal based on the desired signal frequency range [59–62].

Generally, the sensors entail three modules, namely, the detection element, transducer, and signal processor. Once the source to be detected contacts the detection element, the detection element changes and converted into a signal by the transducer. This signal is then processed by the signal processor. In order to select the suitable sensor, the detection element is considered the most significant component. Thus, for functional biosensor, the sensor should have a discerning detection element according to the human body signal to be detected. Then, the transducers can be thermal, piezoelectric, optical, and electrochemical transducers. Finally, the signal processing can be a meter or a simple circuit to control the data acquisition.

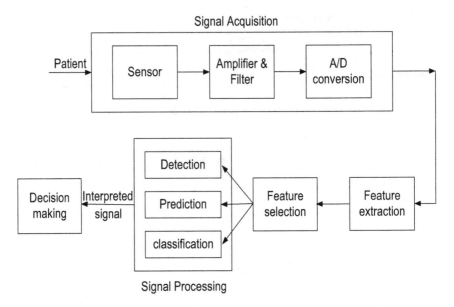

Fig. 2.6 Bioinformation acquisition and processing phases

2.3 Biomedical Signal Acquisition

Biosignals are tiny and contain unwanted interference or noise that obscures the relevant information in the measured biosignal. Thus, refined bioinformation acquisition equipments/methods are used to extract the significant information from the biomedical signals. The essential components of the biomedical signal acquisition systems include data acquisition, amplifiers, sensors, analog signal conditioner, digital signal processing circuit, and bioinformation data storage as demonstrated in Fig. 2.6.

Figure 2.6 includes the sensors that detect the biosignal under observation that is amplified and then converted into digital biosignal to adapt the data acquisition system requirements and for storage and decision making. For noise reduction and significant bioinformation extraction, digital signal processing methods are conducted to improve the physiological understanding from the captured biosignal. These phases are used in almost all medical instruments, where the employed sensors selected according to the type of the captured biosignal. One of the most essential permanent biosignals is the acoustic biosignal that creates internally in the human body, such as the snoring sounds, the lung sounds, and the heart sounds, that arise in the course of the vital functions of the body and provides physiological [63–67]. Typically, in the human body, vibrating structures produce acoustic sounds, which are damped during their propagating through the thoracic tissues to the skin. Dissimilar body sounds interfere at the skin and make the mechanical vibration of the skin that can be observed by sound sensors, which convert the acoustic biosignal into electric signals. Different interdisciplinary acoustic biosignal nature and acoustic sensors are challenging to cope with the technology advancement.

References

1. Marchionini, G. (1997). *Information seeking in electronic environments* (Vol. 9). New York: Cambridge University Press.
2. Yilmaz, T., Foster, R., & Hao, Y. (2010). Detecting vital signs with wearable wireless sensors. *Sensors, 10*(12), 10837–10862.
3. Dey, N., & Ashour, A. S. (2018). Sources localization and DOAE techniques of moving multiple sources. In *Direction of arrival estimation and localization of multi-speech sources* (pp. 23–34). Cham: Springer.
4. Dey, N., & Ashour, A. S. (2018). Computing in medical image analysis. In *Soft computing based medical image analysis* (pp. 3–11).
5. Elhayatmy, G., Dey, N., & Ashour, A. S. (2018). Internet of things based wireless body area network in healthcare. In *Internet of things and big data analytics toward next-generation intelligence* (pp. 3–20). Cham: Springer.
6. Ghaderi, F. (2010). *Signal processing techniques for extracting signals with periodic structure: Applications to biomedical signals*. Cardiff University.
7. Odinaka, I. C. (2014). *Identifying humans by the shape of their heartbeats and materials by their X-ray scattering profiles*. Washington University in St. Louis.
8. Haraldsson, H., Edenbrandt, L., & Ohlsson, M. (2004). Detecting acute myocardial infarction in the 12-lead ECG using Hermite expansions and neural networks. *Artificial Intelligence in Medicine, 32*(2), 127–136.
9. Dey, N., & Ashour, A. S. (2018). Applied examples and applications of localization and tracking problem of multiple speech sources. In *Direction of arrival estimation and localization of multi-speech sources* (pp. 35–48). Cham: Springer.
10. Jiminez Gonzalez, A. (2010). *Antenatal foetal monitoring through abdominal phonogram recordings: A single-channel independent component approach* (Doctoral dissertation, University of Southampton).
11. Dickhaus, H., & Heinrich, H. (1996). Classifying biosignals with wavelet networks [a method for noninvasive diagnosis]. *IEEE Engineering in Medicine and Biology Magazine, 15*(5), 103–111.
12. Mar, T., Zaunseder, S., Martínez, J. P., Llamedo, M., & Poll, R. (2011). Optimization of ECG classification by means of feature selection. *IEEE Transactions on Biomedical Engineering, 58*(8), 2168–2177.
13. Tavakolian, K., Nasrabadi, A. M., & Rezaei, S. (2004, May). Selecting better EEG channels for classification of mental tasks. In *Circuits and Systems, 2004. ISCAS'04. Proceedings of the 2004 International Symposium on* (Vol. 3, pp. III–537). IEEE.
14. Arvaneh, M., Guan, C., Ang, K. K., & Quek, C. (2011). Optimizing the channel selection and classification accuracy in EEG-based BCI. *IEEE Transactions on Biomedical Engineering, 58*(6), 1865–1873.
15. Martínez-Vargas, J. D., Godino-Llorente, J. I., & Castellanos-Dominguez, G. (2012). Time–frequency based feature selection for discrimination of non-stationary biosignals. *EURASIP Journal on Advances in Signal Processing, 2012*(1), 219.
16. Vidaurre, C., Sander, T. H., & Schlögl, A. (2011). BioSig: The free and open source software library for biomedical signal processing. *Computational Intelligence and Neuroscience, 2011*.
17. Zhang, Z., Song, Y., Cui, H., Wu, J., Schwartz, F., & Qi, H. (2017). Topological analysis and Gaussian decision tree: Effective representation and classification of biosignals of small sample size. *IEEE Transactions on Biomedical Engineering, 64*(9), 2288–2299.
18. Georgieva, O., Milanov, S., & Georgieva, P. (2014). Unsupervised EEG biosignal discrimination. *International Journal of Reasoning-based Intelligent Systems, 6*(3–4), 118–125.
19. Cuesta-Frau, D., Pérez-Cortes, J. C., Andreu-García, G., & Novák, D. (2002). Feature extraction methods applied to the clustering of electrocardiographic signals. A comparative study. In *Pattern Recognition, 2002. Proceedings. 16th International Conference on* (Vol. 3, pp. 961–964). IEEE.

20. Kim, J., Mastnik, S., & André, E. (2008, January). EMG-based hand gesture recognition for realtime biosignal interfacing. In *Proceedings of the 13th international conference on Intelligent user interfaces* (pp. 30–39). ACM.

21. Prutchi, D., & Norris, M. (2005). *Design and development of medical electronic instrumentation: A practical perspective of the design, construction, and test of medical devices*. Hoboken: Wiley.

22. Kramme, R., Hoffmann, K. P., & Pozos, R. S. (Eds.). (2011). *Springer handbook of medical technology*. New York: Springer Science & Business Media.

23. Bronzino, J. D. (2006). Biomedical signals: Origin and dynamic characteristics; frequency-domain analysis. In *Medical devices and systems* (pp. 27–48). CRC Press.

24. Bronzino, J. D. (Ed.). (2006). *Medical devices and systems*. Boca Raton: CRC Press.

25. Berntson, G. G., Quigley, K. S., & Lozano, D. (2007). Cardiovascular psychophysiology. In J. T. Cacioppo, L. G. Tassinary, & G. G. Berntson (Eds.), *Handbook of psychophysiology* (Vol. 3, pp. 182–210). Cambridge: Cambridge University Press.

26. Casaccia, S., Sirevaag, E. J., Richter, E., O'Sullivan, J. A., Scalise, L., & Rohrbaugh, J. W. (2014, May). Decoding carotid pressure waveforms recorded by laser Doppler vibrometry: Effects of rebreathing. In *AIP Conference Proceedings* (Vol. 1600, No. 1, pp. 298–312). AIP.

27. Soleymani, S., Borzage, M., Noori, S., & Seri, I. (2012). Neonatal hemodynamics: Monitoring, data acquisition and analysis. *Expert Review of Medical Devices, 9*(5), 501–511.

28. Kaniusas, E. (2015). *Biomedical signals and sensors II*. Berlin\Heidelberg: Springer.

29. Liu, Y., Norton, J. J., Qazi, R., Zou, Z., Ammann, K. R., Liu, H., Yan, L., Tran, P. L., Jang, K., Lee, J. W., Zhang, D., Kilian, K. A., Jung, S. H., Bretl, T., Xiao, J., Slepian, M. J., Huang, Y., Jeong, J., & Rogers, J. A. (2016). Epidermal mechano-acoustic sensing electronics for cardiovascular diagnostics and human-machine interfaces. *Science Advances, 2*(11), e1601185.

30. Alamdari, N. T. (2016). *A morphological approach to identify respiratory phases of seismocardiogram*. The University of North Dakota.

31. Fay, C. (2013). *Investigation into strategies for harvesting chemical based information using digital imaging and infra-red sensors for environmental and health applications* (Doctoral dissertation, Dublin City University).

32. Kaniusas, E. (2012). Fundamentals of biosignals. In *Biomedical signals and sensors I* (pp. 1–26). Berlin\Heidelberg: Springer.

33. Safieddine, D., Kachenoura, A., Albera, L., Birot, G., Karfoul, A., Pasnicu, A., Biraben, A., Wendling, F., Senhadji, L., & Merlet, I. (2012). Removal of muscle artifact from EEG data: Comparison between stochastic (ICA and CCA) and deterministic (EMD and wavelet-based) approaches. *EURASIP Journal on Advances in Signal Processing, 2012*(1), 127.

34. Sontakay, R. (2018). *Real-time signal analysis of the ECG signal for generating an artificial pulse for continuous flow blood pumps using virtual instrumentation* (Doctoral dissertation, California State University, Northridge).

35. Dey, N., & Ashour, A. S. (2017). *Direction of arrival estimation and localization of multi-speech sources*. Springer Science and Business Media.

36. Ashour, A. S., Dey, N., & Mohamed, W. S. (2016). Abdominal imaging in clinical applications: Computer aided diagnosis approaches. In *Medical imaging in clinical applications* (pp. 3–17). Cham: Springer.

37. Dey, N., Hassanien, A. E., Bhatt, C., Ashour, A., & Satapathy, S. C. (Eds.). (2018). *Internet of things and big data analytics toward next-generation intelligence*. Cham: Springer.

38. Gospodinova, E., Gospodinov, M., Dey, N., Domuschiev, I., Ashour, A. S., Balas, S. V., & Olariu, T. (2016, August). Specialized software system for heart rate variability analysis: An implementation of nonlinear graphical methods. In *International workshop soft computing applications* (pp. 367–374). Cham: Springer.

39. Acharya, U. R., Joseph, K. P., Kannathal, N., Lim, C. M., & Suri, J. S. (2006). Heart rate variability: a review. *Medical and biological engineering and computing, 44*(12), 1031–1051.

40. Soni, Y., Jain, J. K., Meena, R. S., & Maheshwari, R. (2017, May). HRV analysis of young adults in pre-meal and post-meal stage. In *Recent Trends in Electronics, Information*

& *Communication Technology (RTEICT), 2017 2nd IEEE International Conference on* (pp. 1125–1129). IEEE.

41. Ronkainen, N. J., Halsall, H. B., & Heineman, W. R. (2010). Electrochemical biosensors. *Chemical Society Reviews, 39*(5), 1747–1763.

42. Scheller, F., Schubert, F., Pfeiffer, D., Hintsche, R., Dransfeld, I., Renneberg, R., Wollenberger, U., Riedel, K., Pavlova, M., Kuhn, M., Muller, H. G., Tan, P., Hoffmann, W., & Movitz, W. (1989). Research and development of biosensors. A review. *Analyst, 114*(6), 653–662.

43. Kriz, D., Ramström, O., & Mosbach, K. (1997). Peer reviewed: Molecular imprinting: New possibilities for sensor technology. *Analytical Chemistry, 69*(11), 345A–349A.

44. Deisingh, A. K., & Thompson, M. (2004). Biosensors for the detection of bacteria. *Canadian Journal of Microbiology, 50*(2), 69–77.

45. Shah, J., & Wilkins, E. (2003). Electrochemical biosensors for detection of biological warfare agents. *Electroanalysis, 15*(3), 157–167.

46. Rodriguez-Mozaz, S., Marco, M. P., de Alda, M. J. L., & Barceló, D. (2004). Biosensors for environmental monitoring of endocrine disruptors: A review article. *Analytical and Bioanalytical Chemistry, 378*(3), 588–598.

47. Saha, K., Agasti, S. S., Kim, C., Li, X., & Rotello, V. M. (2012). Gold nanoparticles in chemical and biological sensing. *Chemical Reviews, 112*(5), 2739–2779.

48. Herold, M., Scepan, J., & Clarke, K. C. (2002). The use of remote sensing and landscape metrics to describe structures and changes in urban land uses. *Environment and Planning A, 34*(8), 1443–1458.

49. Mac Ruairí, R., Keane, M. T., & Coleman, G. (2008, August). A wireless sensor network application requirements taxonomy. In *Sensor Technologies and Applications, 2008. SENSORCOMM'08. Second International Conference on* (pp. 209–216). IEEE.

50. Gnawali, O., Yarvis, M., Heidemann, J., & Govindan, R. (2004, October). Interaction of retransmission, blacklisting, and routing metrics for reliability in sensor network routing. In *Sensor and Ad Hoc Communications and Networks, 2004. IEEE SECON 2004. 2004 First Annual IEEE Communications Society Conference on* (pp. 34–43). IEEE.

51. Boccippio, D. J., Koshak, W., Blakeslee, R., Driscoll, K., Mach, D., Buechler, D., Boeck, W., Christian, H. J., & Goodman, S. J. (2000). The Optical Transient Detector (OTD): Instrument characteristics and cross-sensor validation. *Journal of Atmospheric and Oceanic Technology, 17*(4), 441–458.

52. Rothrock, R. L., & Drummond, O. E. (2000, July). Performance metrics for multiple-sensor multiple-target tracking. In *Signal and Data Processing of Small Targets 2000* (Vol. 4048, pp. 521–532). International Society for Optics and Photonics.

53. Nagel, J. H. (2000). Biopotential amplifiers. In J. D. Bronzino (Ed.), *Biomedical engineering hand book* (2nd ed., pp. 70–71). New York: Springer-Verlag.

54. Zhou, G., Wang, Y., & Cui, L. (2015). Biomedical sensor, device and measurement systems. In *Advances in Bioengineering*. InTech.

55. Denison, T. J., Jensen, R. M., & Santa, W. A. (2009). *U.S. Patent Application No. 12/237,868.*

56. Ljubisavljevic, M., & Popovic, M. B. (1999). Data acquisition, processing and storage. In *Modern techniques in neuroscience research* (pp. 1277–1311). Berlin, Heidelberg: Springer.

57. Schreiner, S. (2014). Medical instruments and devices. In *Medical devices and human engineering*.

58. Zikov, T., Bibian, S., & Modarres, M. (2017). *U.S. Patent No. 9,554,721.* Washington, DC: U.S. Patent and Trademark Office.

59. Estrada, E. F. (2010). *Computer-aided detection of sleep apnea and sleep stage classification using HRV and EEG signals.* The University of Texas at El Paso.

60. Cavazzana, L. (2012). Integrating an EMG signal classifier and a hand rehabilitation device: Early signal recognition and real time performances.

61. Rodrigues, F. M. S. (2015). *Establishing a framework for the development of multimodal virtual reality interfaces with applicability in education and clinical practice* (Doctoral dissertation).

62. Estrada, E. F. (2010). *Computer-aided detection of sleep apnea and sleep stage classification using HRV and EEG signals*. The University of Texas at El Paso.
63. Kaniusas, E. (2015). Sensing by acoustic biosignals. In *Biomedical signals and sensors II* (pp. 1–90). Berlin, Heidelberg: Springer.
64. Bridger, K., Cooke, A. V., Kuhn, P. M., Lutian, J. J., Passaro, E. J., Sewell, J. M., Waskey, T. V., & Rubin, G. R. (2002). *U.S. Patent No. 6,491,647*. Washington, DC: U.S. Patent and Trademark Office.
65. Sörnmo, L., & Laguna, P. (2005). *Bioelectrical signal processing in cardiac and neurological applications* (Vol. 8). London: Academic Press.
66. Kaniusas, E. (2007). Acoustical signals of biomechanical systems. In C. T. Leondes (Ed.), *Biomechanical systems technology: Volume 4: General anatomy* (pp. 1–44). Singapore: World Scientific Publishing.
67. Shimizu, K., Kawamura, K., & Yamamoto, K. (2000). Practical considerations for a system to locate moving persons. *Biotelemetry, 15*, 639–645.

Chapter 3
Acoustic Wave Technology

Abstract Sound is the generalized name of the acoustic waves that have frequencies within the range of one to tens of thousands Hertz, where the maximum human hearing ability is 20 kHz. The main role of the sound sensors/transducers is to use electrical energy for creating mechanical vibrations that disturb the surrounding air to produce sound at the inaudible or audible frequencies, which requires a transmission medium. The sound waveform can be characterized by the velocity (m/s), the frequency (f), and the wavelength (λ), like the electrical waveform. The sounds wave shape and frequency are determined by the vibration/origin that created the sound, while the velocity depends on the sound wave transmission. Discovery of the quartz resonator to stabilize the electronic oscillators leads to the detection of the piezoelectricity. Piezoelectricity can be defined as the electrical charges production by the mechanical stress imposition. This creates a revolution in the acoustic wave sensors and devices using a piezoelectric material for generating acoustic waves. Applying a fluctuating electric field by the piezoelectric acoustic wave sensors, a mechanical wave is created that propagates via the substrate and transformed to electric field for further measurements. This chapter reveals about the fundamentals of the acoustics with a detailed explanation of the several body acoustic sounds sources.

3.1 Fundamentals of Acoustics and Psychoacoustics

Sound is the produced wave by vibrating entities. It travels via a medium from one point to another one. It is a mechanical wave as it produced due to the traveling motion of the sound vibration via a non-vacuous (conductive) medium where the mechanical sound wave travels. It results from a longitudinal motion of the medium's particles. The wave's physics explains the process of the sound generation, traveling, and reception, where the sound waves carry the vibration (disturbance) from one position to another that originated from the wave source. The initiating

© The Author(s), under exclusive licence to Springer International Publishing
AG, part of Springer Nature 2019
N. Dey et al., *Acoustic Sensors for Biomedical Applications*,
SpringerBriefs in Electrical and Computer Engineering,
https://doi.org/10.1007/978-3-319-92225-6_3

source may be a stereo speaker or vocal chords. The particle interaction leads to the sound traveling, which lets the vibrating waves to transport from one position to another [1–17]. Conversely, the mechanical sound waves require a receiver to widespread. The general characteristics are required for the acoustic sound waves to be transmitted:

- The acoustic waves are mechanical waves that need a medium to carry their energy from any position to another.
- The acoustic waves, which are mechanical waves, are incapable to travel over a vacuum.
- The acoustic waves are longitudinal waves, which consist of repeating rarefactions/compressions patterns.

There are several methods to measure the sound waves, namely, the (i) frequency; (ii) wavelength, where the distance that the disturbance travels via the medium represents a complete wave cycle; (iii) amplitude which is related to the sound volume, loudness, and intensity; (iv) phase; and (v) speed of sound that depends on the medium state/type, which is affected by the elasticity and the inertia.

Consequently, the sound represents the wave motion with different pressure due to a vibrating source that sets particles in motion only one sound tone. The discrete particles travel about their relaxing point at the same tone frequency. Vibrating particles during their movement push adjacent other ones and put them in motion. This creates a chain effect producing areas of low and high pressure. The interchange between the high and low pressure areas transfers away from the sound source generating sound waves. On a membrane, the mechanical effect is used to sense the sound waves, such as the membrane of a microphone or a diaphragm of a stethoscope. For a real-world example, say there is a trumpet playing in the room [18–20].

The actual sound may be an acoustic wave from a single sound pulse, mechanical vibration, noise, or a continuous frequency sound wave. Audio sound sensors (transducers) include input sensors to transform the sound waves to electrical signal and output actuators to transform back into the electrical signals to sound. Acoustic (sound) sensors can detect and transmit vibrations and sound waves from infrasound (very low frequencies) up to ultrasound (very high frequencies). Acoustic wave sensors detect acoustic or mechanical waves produced by the human body. During the propagation of the acoustic wave through the body, the propagation path characteristics change, which affect the amplitude/velocity of the acoustic wave. Measuring the phase/frequency characteristics of sensed signals reflects the occurred changes in the velocity, which is correlated to the consistent physical measured quantity. Several expressive biosignals are carried by the body sounds to state the patient's health. The mechanical waves within the body generate the body sounds due to the mechanical vibrations of blood/tissues, airway walls oscillation, and the heart valve vibrations. These body sounds composed of several spectral varying frequency/intensity components due to the existence of the noises. The heart sounds auscultation is useful for detecting the cardiac pathologies, while the snoring/lung sounds auscultation are used for detecting the respiratory disorders [21–26].

Fig. 3.1 Electrical circuit model of the acoustic biosignal formation and sensing processes

Figure 3.1 illustrates the electrical circuit model of the body sound formation and sensing phases of the acoustic biosignals.

The formation of the body sounds includes their origination and transmission through the tissue. The body sounds acoustical path initiates at the sound source, which vibrates the volumes of the blood, and oscillates the biological structures. In the biological medium, the body sound propagates with velocity v, which equals the sound propagation. In addition, in the time domain, the body sound oscillates with the sound frequency f. It also oscillates with wavelength λ along its propagation path.

3.2 Acoustic Biosignal Sources

There are several body sounds that lead to acoustic biosignals.

3.2.1 Heart Sounds

The cardiac system's contractile activity influences the heart sounds (HS), which produce direct information on the closure of the heart's valves. The HSs provide information on myocardial, hemodynamic, and valvular activities weakening. The normal heart sounds include the sounds related to the atrioventricular valves closing to prevent the blood backward flow. Induced mechanical vibrations are obvious as the first HS due to blood flow deceleration, ventricular myocardium jerky contraction, and any abrupt tension changes. This first HS is the longest and loudest compared to all other HSs. It includes relatively low-frequency spectral components with duration of about 140 ms. The semilunar valves closing produce the second HS, which prevents the blood backward flow. In the second HS, during the inspiration, the left-sided sounds lead by about 40 ms, while with expiration both the right-sided/left-sided sounds are overlaid or still marginally split by <30 ms [27–32].

Figure 3.2 illustrates the different heart's sound waveforms. The normal first HS is louder than the second HS in the mitral valve [33–38]. The normal minimally split first HS is a normal variation of the first HS (Fig. 3.2b). Abnormal right bundle branch block can be detected if the first heart sound splitting takes>50 milliseconds.

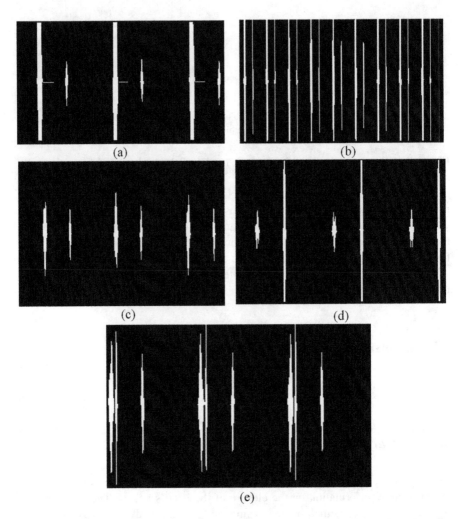

Fig. 3.2 Heart sounds different waveform (**a**) normal first heart sound, (**b**) normal minimally split first heart sound, (**c**) abnormal markedly split first heart sound, (**d**) decreased intensity first heart sound, and (**e**) abnormal first heart sound plus aortic ejection click

In the tricuspid area, the splitting is obviously heard. In addition, due to several heart abnormalities, a decreased intensity first HS can be produced as shown in Fig. 3.2d. Figure 3.2e shows an aortic ejection click abnormality caused by stiffness and thickened of the aortic valve cusps.

The second HS has about 110 ms duration with more snapping quality, higher frequency components, and lower intensity compared to the first HS. Other various HSs occur based on the existence of abnormality or the age effect, such as:

(i) The HS due to the rapid filling of the ventricle, which is comparatively short and comprises very low-frequency components of 25–50 Hz range.

(ii) The HS due to the active ventricular filling and the atrial contraction. This sound contains 20–30 Hz very low-frequency spectral components.

(iii) The ejection sounds due to the semilunar valve opening, where a HS is generated from the abrupt valves opening or the occurrence of sudden tensing leading to the clicky sounds of high frequency.

(iv) The opening sounds due to the atrioventricular valves opening.

(v) The murmurs are abnormal sounds due to the convinced turbulent blood flow in the backward regurgitation progress, which are high-frequency noisy sounds.

Figure 3.2 illustrates the different waveforms of the heart sounds.

3.2.2 Breath Sound

The lungs and the large airways produce in/out breath sounds (BSs) that can be received by stethoscope, which can be normal or abnormal. Lung sounds provide information about the ventilation/dynamics of the upper airways. Abnormal BSs specify a lung problem, including infection, inflammation, asthma, obstruction, and fluid in the lungs. Identifying such medical conditions requires listening to the BSs. The lung sounds' time amplitude plots can be represented by expanding or unexpanding ways, where the expanded time scales (at 800 mm/sec) illustrate distinct patterns not appearing in the unexpanded speed plots (at 100 mm/sec) as demonstrated in Fig. 3.3 [39–43]. The right waveforms in Fig. 3.3 represent the expanded form, while the left waveforms represent the unexpanded form, where the amplitude and time are represented on the Y- and X-axis, respectively.

Figure 3.3 shows more details of the acoustic phenomena in the time expanded analysis waveforms compared to the unexpanded ones as they are stretched out to provide an overall view of the acoustic characteristics of the inspiratory sounds in the real time. However, the unexpanded display is similar to the phonocardiographic presentation. Figure 3.3 illustrates different normal and abnormal inspiration sound waveforms. Generally, the lung sounds cannot be recognized clearly in the time domain, while the heart sounds can be easily observed due to their comparatively high intensity. The HSs are about 30 dB stronger compared to the inspiratory sounds when auscultated on the chest. The tracheobronchial normal sounds happen during both the inspiration and expiration processes, although the vesicular sounds dominate during the inspiration process only. Figure 3.3 reveals that the tracheobronchial sounds have wider frequencies which range up to 1 kHz compared to the vesicular sounds that have components up to 500 Hz. Typically, the normal breathing sound is similar to the air sound, which can be categorized according to the sound source location as follows [44–47]:

(i) Tracheobronchial sounds are originated in the tracheal and bronchial tracts and heard close to the large airways. They are dominated on the neck. The source of this sound is given by the air turbulence that flow in the trachea and bronchi

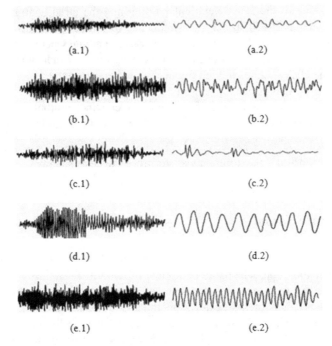

Fig. 3.3 Inspiratory sounds: (*a*) normal vesicular inspiratory sound, (*b*) normal tracheal inspiration, (*c*) abnormal inspiratory crackles, (*d*) abnormal sonorous rhonchus, and (*e*) abnormal inspiratory wheeze

due to high airflow velocity causing air vibrations. The sounds' propagation distance near the skin is comparatively short. Thus, the produced sounds are relatively loud similar to the air blown over a tube and hold up to 1 kHz frequency components, where the amplitude component at the baseline has frequency 1.2–1.8 kHz.

(ii) Vesicular sounds are heard distant positions from the large airways, where the sources of this sounds are spread through the lungs and create in air turbulences. They are dominated at the chest. During the inspiration, the vesicular sounds are originated mainly when the air moves through gradually smaller airways, where the inspiratory airflow hits the airway branches creating the air turbulences. Directional fluctuations of the local airflow occur due to the air turbulences that convinced by branching and bronchospasm of the airways. Nevertheless, the air movements through gradually larger airways occur during the expiration leading to less turbulence and hence less sound during the expiration. The propagation of the vesicular sounds through the lung toward the skin experiences a comparatively large damping, which emits soft sound. The vesicular sounds contain frequency components in the range of 100–400 Hz and mainly at about 100 Hz. The vesicular sounds have narrower spectral range and lower intensity compared to the tracheobronchial sounds.

(iii) Bronchovesicular sounds have intermediate characteristics between the vesicular and tracheobronchial sounds.

However, the abnormal BSs may include a (i) high-pitched BS (Crackles), (ii) low-pitched BS (rhonchi), (iii) vibratory sound due to the contraction of the upper airway (stridor), and (iv) high-pitched whistling sound due to the bronchospasm of the bronchial tubes (wheezing).

3.2.3 Gurgling/Intestinal Sound

Auscultation abdominal sounds are produced by blood flow, friction rubs, and peristalsis. The intestine produces abdominal sounds known as the rumbling, gurgling, high-pitched, and growling sounds. These sounds are related to the movement of liquids, food, juices, digestive process, and air through the intestines. The peristalsis causes rumbling sound after eating. Hunger sends signals through the brain to the intestines and stomach resulting in the muscle contraction causing sounds. Abdominal sounds can be categorized as normal, hyperactive, or hypoactive [48–51]. Hyperactive bowel sounds are louder sounds than the normal that indicate the increased intestinal activity. Conversely, the hypoactive bowel sounds occur with the slowed down intestinal activity. Hypoactive, hyperactive, or absent bowel sounds may indicate one of the following diseases: digestive tract infection, trauma, hernia, less blood flow to the intestines, tumor, abnormal potassium/calcium level in the blood, perforated ulcers, and intestinal movement temporary reducing.

The stethoscope is used to hear any abnormal bowel sounds, which is called auscultation [52–54]. However, bowel obstructions yield high-pitched, very loud sounds. These sounds can often be heard without using a stethoscope. Bowel sounds are variable; thus the stethoscope diaphragm is used mainly to hear the bowel sounds to note their character and frequency, where the gurgling sounds occur at a frequency of 5–34 per min. Abnormal hyperactive bowel sounds have high-pitched and increased bowel sounds.

3.2.4 Korotkoff Sounds

The Korotkoff sounds are the sounds heard during the blood pressure measurements using the stethoscope. These sounds are different than the heart sounds "dub" and "lub" as they are due to vibrations inside the ventricles during the valves' snapping shut, which are heard with the stethoscope. The first Korotkoff sound is heard if the pressure in the cuff decreases to the level of the patient's systolic blood pressure level produced by the heart due to the occurred turbulence. Thumping sounds have been heard since the pressure in the cuff is permitted to fall more. Ultimately, the sounds' quality changes as the pressure in the cuff falls more till they disappear, where decreasing the pressure lower than the diastolic blood pressure will cancel the control of the cuff on the blood flow. This returns the blood flow to be again smooth without turbulence leading to no more audible sound [55–57].

3.2.5   Other Body Sounds

3.2.5.1   Vascular Sounds

Audible vascular (bruits) sounds are heard due to the turbulent of the blood flow in the large arteries, namely, the iliac, renal arteries, femoral arteries, and aorta. These sounds can be heard at different vascular locations for at least 5 sec each. Swishing sounds may be produced during the auscultation bruits indicating abdominal aortic aneurism, iliac/femoral artery stenosis, and renal artery stenosis [58].

3.2.5.2   Friction Rub Sounds

The stethoscope is used to hear the friction rubs over the liver and spleen. Friction rub indicates peritoneal surface's inflammation of the organ due to tumor, infarct, or infection. Generally, from the preceding reporting of the human body sounds, we can conclude that the airways and lungs require different instruments/sensors to detect them than those used to hear the heart sounds. The stethoscope is used to detect such sounds, which is placed over the chest while breathing in/out slowly and deeply. Through the bell in the stethoscope, the listener will hear different sounds at the different positions of the chest. Afterward, in the same way, the diaphragm is used. In normal lung sounds, there will be no crackles or wheezes. Crackles are heard when the lung rubs against the chest wall, creating friction and rubbing sound. Wheeze is a whistling, high-pitched sound that is heard with the constricted airways with the existence of fluid in the lungs. In addition, the stethoscope is used to hear by the small intestines and stomach gurgling/intestinal sounds when placed over the abdomen upper left part below the ribs. However, the borborygmi noise due to the movement of the fecal material, gas, or food is also heard [59–61].

The overall human hearing ranges from 20 to 20,000 Hz. The human HSs have a frequency within the 20–200 Hz range, and the human lung sounds have a frequency range of 25–1500 Hz. Acoustic wave sensors are essential to sense such frequencies, especially with the existence of the different sources of noise [61–65]. These sensors are sensitive with varying levels to the alarms/changes from many physical parameters. The advancement in the electronics and sensor/transducer fabrication leads to rapid progress in the acoustic medical devices.

References

1. Major, F. G. (2014). The quartz revolution. In *Quo Vadis: Evolution of modern navigation* (pp. 131–150). New York: Springer.
2. McGahey, C. S. (2009). *Harnessing nature's timekeeper: A history of the piezoelectric quartz crystal technological community (1880–1959)* (Doctoral dissertation, Georgia Institute of Technology).

3. Major, F. G. (2013). *The quantum beat: The physical principles of atomic clocks.* Springer Science & Business Media.
4. Banica, F. G. (2012). *Chemical sensors and biosensors: Fundamentals and applications.* Chichester: Wiley.
5. Park, N. (2015). *Development of integration of sensors and circuits for wearable electronics.* San Diego: University of California.
6. Vorländer, M. (2007). *Auralization: Fundamentals of acoustics, modelling, simulation, algorithms and acoustic virtual reality.* Berlin: Springer Science & Business Media.
7. Hansen, C. H. (2001). Fundamentals of acoustics. In *Occupational exposure to noise: Evaluation, prevention and control* (pp. 23–52). Geneva: World Health Organization.
8. Jacobsen, F., Poulsen, T., Rindel, J. H., Gade, A. C., & Ohlrich, M. (2011). *Fundamentals of acoustics and noise control.* Department of Electrical Engineering, Technical University of Denmark.
9. Baumgarte, F., & Faller, C. (2003). Binaural cue coding-Part I: Psychoacoustic fundamentals and design principles. *IEEE Transactions on Speech and Audio Processing, 11*(6), 509–519.
10. Weisenberger, J. M. (1994). Fundamentals of Hearing: An Introduction. *Ear and Hearing, 15*(5), 409.
11. Müller, G., & Möser, M. (Eds.). (2012). *Handbook of engineering acoustics.* Springer Science & Business Media.
12. Toole, F. E. (2000, September). The acoustics and psychoacoustics of loudspeakers and rooms-the stereo past and the multichannel future. In *Audio engineering society convention 109.* Audio Engineering Society.
13. Zwicker, E., & Fastl, H. (2013). *Psychoacoustics: Facts and models* (Vol. 22). Springer Science & Business Media.
14. Raichel, D. R. (2006). *The science and applications of acoustics.* Springer Science & Business Media.
15. Schleske, M. (2002). Empirical tools in contemporary violin making: Part II. Psychoacoustic analysis and use of acoustical tools. *CAS Journal, 4*(6), 43.
16. Lundén, P., Gustin, M., Nilsson, M. E., Forssén, J., & Hellström, B. (2010, September). Psychoacoustic evaluation as a tool for optimization in the development of an urban soundscape simulator. In *Proceedings of the 5th Audio Mostly Conference: A Conference on Interaction with Sound* (p. 3). ACM.
17. Blauert, J. (Ed.). (2005). *Communication acoustics* (Vol. 2). Berlin: Springer.
18. Sterne, J. (2003). *The audible past: Cultural origins of sound reproduction.* Durham\London: Duke University Press.
19. Hollerrweger, F. (2011). *The revolution is hear!: Sound art, the everyday and aural awareness* (Doctoral dissertation, School of Music and Sonic Arts, Faculty of Arts, Humanities and Social Sciences, Queen's University Belfast).
20. Tichindeleanu, O. (2008). *The graphic sound. An archeology of sound, technology and knowledge at 1900* (Doctoral dissertation, State University of New York at Binghamton).
21. Tavel, M. E. (2006). Cardiac auscultation: A glorious past—And it does have a future! *Circulation, 113*(9), 1255–1259.
22. Thompson, W. R., Hayek, C. S., Tuchinda, C., Telford, J. K., & Lombardo, J. S. (2001). Automated cardiac auscultation for detection of pathologic heart murmurs. *Pediatric Cardiology, 22*(5), 373–379.
23. Jiang, Z., & Choi, S. (2006). A cardiac sound characteristic waveform method for in-home heart disorder monitoring with electric stethoscope. *Expert Systems with Applications, 31*(2), 286–298.
24. Reed, T. R., Reed, N. E., & Fritzson, P. (2004). Heart sound analysis for symptom detection and computer-aided diagnosis. *Simulation Modelling Practice and Theory, 12*(2), 129–146.
25. Akay, M., Semmlow, J. L., Welkowitz, W., Bauer, M. D., & Kostis, J. B. (1990). Detection of coronary occlusions using autoregressive modeling of diastolic heart sounds. *IEEE Transactions on Biomedical Engineering, 37*(4), 366–373.

26. Ari, S., Hembram, K., & Saha, G. (2010). Detection of cardiac abnormality from PCG signal using LMS based least square SVM classifier. *Expert Systems with Applications, 37*(12), 8019–8026.

27. Wood, J. C., & Barry, D. T. (1996). Time-frequency analysis of skeletal muscle and cardiac vibrations. *Proceedings of the IEEE, 84*(9), 1281–1294.

28. Seo, J. H., & Mittal, R. (2011). A high-order immersed boundary method for acoustic wave scattering and low-Mach number flow-induced sound in complex geometries. *Journal of Computational Physics, 230*(4), 1000–1019.

29. Ibarra-Hernández, R. F., Alonso-Arévalo, M. A., Cruz-Gutiérrez, A., Licona-Chávez, A. L., & Villarreal-Reyes, S. (2017). Design and evaluation of a parametric model for cardiac sounds. *Computers in Biology and Medicine, 89*, 170–180.

30. Kong, Y., Jenkinson, M., Andersson, J., Tracey, I., & Brooks, J. C. (2012). Assessment of physiological noise modelling methods for functional imaging of the spinal cord. *NeuroImage, 60*(2), 1538–1549.

31. Ryu, K., Keel, A., Hellman, H., & Svensson, T. (2013). *U.S. Patent No. 8,583,230.* Washington, DC: U.S. Patent and Trademark Office.

32. Gayeski, N., McMillan, B., & Trotter, P. (2011). Historical abundance of Puget Sound steelhead, Oncorhynchus mykiss, estimated from catch record data. *Canadian Journal of Fisheries and Aquatic Sciences, 68*(3), 498–510.

33. Malone, D. J. (2006). Cardiovascular diseases and disorders. In *Physical therapy in acute care: A clinician's guide* (pp. 139–209). Thorofare, NJ: Slack.

34. Rabinovitch, M. (2011). 5 pulmonary vascular pathophysiology. *Pediatric Cardiovascular Medicine*.

35. Rosen, D. A., & Rosen, K. R. (2005). Anomalies of the aortic arch and valve. In *Pediatric cardiac anesthesia* (pp. 381–424). Philadelphia: Lippincott Williams & Wilkins.

36. Silberbach, M., & Hannon, D. (2007). Presentation of congenital heart disease in the neonate and young infant. *Pediatrics in Review, 28*(4), 123.

37. Bergeron, B. P., & Greenes, R. A. (1989). Clinical skill-building simulations in cardiology: HeartLab and EkgLab. *Computer Methods and Programs in Biomedicine, 30*(2–3), 111–126.

38. Bergeron, B. P. (1988). HeartLab: A multi-mode simulation for teaching cardiac auscultation. *SIMULATION, 50*(2), 50–60.

39. Ducharme, F. M., Sze, M. T., & Chauhan, B. (2014). Diagnosis, management, and prognosis of preschool wheeze. *The Lancet, 383*(9928), 1593–1604.

40. Everard, M. L. (2006). The relationship between respiratory syncytial virus infections and the development of wheezing and asthma in children. *Current Opinion in Allergy and Clinical Immunology, 6*(1), 56–61.

41. Schreur, H. J., Vanderschoot, J., Zwinderman, A. H., Dijkman, J. H., & Sterk, P. J. (1994). Abnormal lung sounds in patients with asthma during episodes with normal lung function. *Chest, 106*(1), 91–99.

42. Bush, A. (2007). Diagnosis of asthma in children under five. *Primary Care Respiratory Journal, 16*(1), 7.

43. Corcoran, B. M., Foster, D. J., & Fuentes, V. L. (1995). Feline asthma syndrome: A retrospective study of the clinical presentation in 29 cats. *Journal of Small Animal Practice, 36*(11), 481–488.

44. Kaniusas, E. (2007). Acoustical signals of biomechanical systems. In *Biomechanical systems technology: Volume 4: General anatomy* (pp. 1–44).

45. Hadjileontiadis, L. J., Rekanos, I. T., & Panas, S. M. (2006). Bioacoustic signals. *Wiley Encyclopedia of Biomedical Engineering*.

46. Kaniusas, E. (2015). Sensing by acoustic biosignals. In *Biomedical signals and sensors II* (pp. 1–90). Berlin\Heidelberg: Springer.

47. Racz, L., Yamamoto, D. P., & Eninger, R. M. (Eds.). (2017). *Handbook of respiratory protection: Safeguarding against current and emerging hazards.* CRC Press.

48. Vanek, V. W., & Al-Salti, M. (1986). Acute pseudo-obstruction of the colon (Ogilvie's syndrome). *Diseases of the Colon & Rectum, 29*(3), 203–210.
49. De Block, C. E., Leeuw, I. H., & Gaal, L. F. (1999). Premenstrual attacks of acute intermittent porphyria: Hormonal and metabolic aspects-a case report. *European Journal of Endocrinology, 141*(1), 50–54.
50. McMillan, S. C. (2004). Assessing and managing opiate-induced constipation in adults with cancer. *Cancer Control, 11*(3_suppl), 3–9.
51. Felder, S., Margel, D., Murrell, Z., & Fleshner, P. (2014). Usefulness of bowel sound auscultation: A prospective evaluation. *Journal of Surgical Education, 71*(5), 768–773.
52. Cannon, W. B. (1905). Auscultation of the rhythmic sounds produced by the stomach and intestines. *American Journal of Physiology-Legacy Content, 14*(4), 339–353.
53. Leng, S., San Tan, R., Chai, K. T. C., Wang, C., Ghista, D., & Zhong, L. (2015). The electronic stethoscope. *Biomedical Engineering Online, 14*(1), 66.
54. Holmgren, C. (1992). Abdominal assessment. *RN, 55*(3), 28–34.
55. Olmstead, C. D., Miller, C. G., Woods, J., & Dhingra, S. (2015). *U.S. Patent Application No. 14/606,487.*
56. Boulpaep, E. L. (2012). Organization of the cardiovascular system. *Medical Physiology,* 429–447.
57. Collins, S. L., Sinsabaugh, R. L., Crenshaw, C., Green, L., Porras-Alfaro, A., Stursova, M., & Zeglin, L. H. (2008). Pulse dynamics and microbial processes in aridland ecosystems. *Journal of Ecology, 96*(3), 413–420.
58. Bacharach, J. M., Biller, J., Fine, L. J., Gray, B. H., Gray, W. A., Gupta, R., Hamburg, N. M., Katzen, B. T., Lookstein, R. A., Lumsden, A. B., Newburger, J. W., Rundek, T., Sperati, C. J., Stanley, J. C., American Heart Association Council on Peripheral Vascular Disease, American Heart Association Council on Clinical Cardiology, American Heart Association Council on Cardiopulmonary, Critical Care, Perioperative and Resuscitation; American Heart Association Council on Cardiovascular Disease in the Young, American Heart Association Council on Cardiovascular Radiology and Intervention; American Heart Association Council on Epidemiology and Prevention, American Heart Association Council on Functional Genomics and Translational Biology, American Heart Association Council for High Blood Pressure Research, American Heart Association Council on the Kidney in Cardiovascular Disease, & American Heart Association Stroke Council. (2014). Fibromuscular dysplasia: State of the science and critical unanswered questions. *Circulation, 129,* 1048–1078.
59. Nagasaka, Y. (2012). Lung sounds in bronchial asthma. *Allergology International, 61*(3), 353–363.
60. Moussavi, Z. (2006). Fundamentals of respiratory sounds and analysis. *Synthesis Lectures on Biomedical Engineering, 1*(1), 1–68.
61. Winland-Brown, J. E., & Klause, K. T. (2017). Respiratory problems. *Family practice and adult-gerontology primary care nurse practitioner certification examination review questions and strategies*, p. 239.
62. Dokur, Z., & Ölmez, T. (2008). Heart sound classification using wavelet transform and incremental self-organizing map. *Digital Signal Processing, 18*(6), 951–959.
63. Naseri, H., & Homaeinezhad, M. R. (2013). Detection and boundary identification of phonocardiogram sounds using an expert frequency-energy based metric. *Annals of Biomedical Engineering, 41*(2), 279–292.
64. Choi, S., & Jiang, Z. (2008). Comparison of envelope extraction algorithms for cardiac sound signal segmentation. *Expert Systems with Applications, 34*(2), 1056–1069.
65. Dokur, Z., & Ölmez, T. (2009). Feature determination for heart sounds based on divergence analysis. *Digital Signal Processing, 19*(3), 521–531.

Chapter 4
Acoustic Sensors

Abstract An acoustic wave biosensor employs mechanical or acoustic waves as a detection instrument to attain biochemical, biophysical, and medical information. It senses changes in elasticity, mass, dielectric properties, and conductivity from the electrical or mechanical variations. At an input transducer, the piezoelectric effect is employed in these devices to stimulate the acoustic waves electrically and to obtain the waves at the output transducer. Acoustic biosensors are implemented with robust piezoelectric crystals such as lithium tantalite, lithium niobate, or quartz that can detect various biomolecules. Sound waves are generated by different compression and expansion of the medium at specific frequencies. For auscultation and listening to body sounds, the stethoscope instrument is used to hear the sounds produced from the heart, intestinal tract, lungs, stomach, the blood flow in the exterior vessels, venous, arterial, uterine, and the sound of human's/animal fetuses. Acoustic wave sensors are convenient in several applications as predominantly mass sensitive devices capable of the respond to small environmental perturbations. New surface acoustic wave devices using different materials for chemical and biological sensing are developed. The improvement of a broad sensor system for biomarker and chemical sensing attracts several researchers. This chapter introduces in details the piezoelectricity effect as well as the acoustic sensor design and the acoustic stethoscope. Finally, the acoustic wave sensors including the bulk acoustic wave sensors, the surface acoustic wave sensors, and the acoustic wave propagation modes are introduced.

4.1 Piezoelectricity Effect

Acoustic sound is a mechanical, longitudinal wave with rarefactions and compressions produced by vibrating the sound source. This motion travels through a non-vacuous medium. In order to quantity the acoustic sound, the frequency, amplitude, speed, wavelength, phase, and time are measured. Accordingly, in order

N. Dey et al., *Acoustic Sensors for Biomedical Applications*,
SpringerBriefs in Electrical and Computer Engineering,
https://doi.org/10.1007/978-3-319-92225-6_4

to measure these parameters of the acoustic (mechanical) sounds, the piezoelectricity defined as the capability of specific materials to generate a voltage due to applying mechanical stress is considered. In addition, the shape of the piezoelectric materials can slightly change with the subject to an external applied voltage.

On the surface of the piezoelectric material, electric charges are produced due to applying any mechanical stress, such as the one applied from the sound waves. In the direct piezoelectric effect, the convinced charges are proportional to the mechanical stress. This piezoelectricity effect has a wide application in detecting the acoustic sound and electronic frequency/high voltage generation. The negative and positive electrical charges are divided in the piezoelectric crystal leading to electrically neutral overall crystal. This symmetry is disturbed with applying a stress to the piezoelectric materials, and the asymmetry of the charge produces a voltage. The piezoelectric material can be categorized according its cutting procedure, namely, shear, longitudinal, and transverse, which are defined as follows [1–8].

(i) Shear effect: due to this effect, the generated charges are independent of the element's shape/size and proportional to the applied forces. The charge is given by

$$C_S = 2d_{xx}A_x m \tag{4.1}$$

where m is the number of elements, which are electrically in parallel and mechanically in series. By applying a force in the x-direction, the piezoelectric coefficient is represented by d_{xx}, where A_x is the applied.

Longitudinal effect: due to this effect, the displaced amount of charge is independent of the piezoelectric element's shape/size and proportional to the applied forces. The released charge due to such configuration can be expressed as

$$C_S = d_{xx}A_x m \tag{4.2}$$

(ii) Transverse effect: due to this effect, the applied force along the y-axis transfers charges along the x-direction, which is perpendicular to the force line. Thus, the generated charge can be expressed by

$$C_x = d_{xy}A_y a / b \tag{4.3}$$

where b is the line dimension and a is in line with the charge generating axis. The transverse effect enables fine-tuning of the element dimension and the applied force.

The piezoelectric materials' status can be piezoceramics, single-crystal materials, piezocomposites, piezofilms, and piezopolymers. The piezoelectric ceramic material is considered the most broadly used type in several applications. However, in surface acoustic devices and frequency-stabilized oscillators, the single-crystal material is still having a significant performance. The polycrystalline materials, such as the barium titanate (BaTiO3), are considered of the most extensively used piezoelectric materials.

Furthermore, the lithium niobate (LiNbO3), the lithium tantalite (LiTaO3), and the quartz are the most prevalent single-crystal piezoelectric materials. The single crystals are anisotropic, presenting dissimilar material characteristics based on the materials' cutoff and the surface wave propagation direction. Moreover, the piezoelectric Pb(Ti,Zr)O3 solid solutions "PZT" ceramics have superior piezoelectric properties. Thus, they are used mainly in the surface acoustic wave device applications.

In piezoelectrics, there are several figures of merit, namely, the piezoelectric voltage constant, the piezoelectric strain constant, the mechanical quality factor, the electromechanical coupling factor, and the acoustic impedance. In surveillance devices, hearing aids, industrial monitoring, and biometrics, the micromachined acoustic sensors that imply the piezoelectric material are used. These sensors can be integrated with on-chip circuits and reduces the acoustic sensors' size compared to the conventional sensors [9]. For surface acoustic wave (SAW), the thin film zinc oxide shows outstanding piezoelectric properties [10]. In order to transform the diaphragm mechanical deflection into a piezoelectric charge distribution, the sputtered piezoelectric ZnO layer can be used. Furthermore, a square silicon diaphragm can be included by anisotropic-oriented silicon wafer etching to fix the silicon membrane's final thickness [11].

4.2 Acoustic Sensor Design

Extensively, the piezoelectricity is used in various devices fabrication, including surface acoustic wave devices, transducers, and actuators. The piezoelectric sensor/transducer can be modeled as a set of filters and a voltage source to represent its very high output impedance. The sensor's voltage source is directly proportional to the applied pressure or force. Then, the output signal is associated with this mechanical force. The effect of the mechanical construction of the sensor is also modeled. The inertia/mass of the sensor as well as an infinite size inertial mass can be represented by an inductance and static capacitance of the sensor, respectively, within the transducer's model [12–14].

In sensors' design, the piezoelectric voltage constant "w" is considered the most important figures of merit, where:

$$E = w\,X \tag{4.4}$$

here E is the induced electric field, and X is the external stress over w [15–17]. In addition, the acoustic impedance parameter is considered in order to evaluate the acoustic energy transfer between two materials, which is given by:

$$M_z^2 = \frac{\text{Pressure}}{\text{Volume velocity}} \tag{4.5}$$

In order to use the piezoelectric material as a sensor, the frequency response's flat region between the resonant peak and the high-pass cutoff is used. The equivalent circuit for the piezoelectric actuator model can be represented by a grouping of R, C, and L.

In surface acoustic wave (SAW) and bulk acoustic devices, the ZnO has thin films, and large piezoelectric coupling is widely used. Ultrasonic waves are now used in various fields. Magnetostrictive materials and piezoelectric ceramics are involved mainly in the design of the sound sources. Piezoceramics are superior in size/efficiency compared to magnetostrictive materials, where a liquid medium is efficient for sound energy transfer.

4.3 Acoustic Stethoscope

Blood flow, friction rubs, and peristalsis produce auscultation abdominal sounds. For auscultation, an acoustic medical device called the stethoscope is used to hear the internal body's sounds. The stethoscope is used as the main tool to diagnose the sounds produced in the abdomen, heart, and thoracic. The conventional stethoscope was monoaural consisting of a chest piece, tube, and earpiece. It transmitted the sound from the chest piece through the tubes to the listener. The chest piece contains two sides: at one of them, a diaphragm is located; and on the other side, a bell exists to listen to the different body sounds. The body sounds vibrate the diaphragm once the diaphragm is placed on the body. These sound waves travel via the tubes to the listener. Instead, if the bell is placed on the body, sound waves are produced due to the vibrations on the skin. Generally, the diaphragm and the bell are used to transmit high- and low-frequency sounds, respectively. Nevertheless, the acoustic stethoscopes' major problem is the effect of the noisy environment on the low intensity sound level.

Afterward, a binaural stethoscope was used, and then the electronic stethoscope was developed to hear the heart sounds. The electronic stethoscope has the similar function of the conventional stethoscope with converting the sound to an electronic signal for further transmutation to the listener through the wire. Its functionalities include the sound signal amplification, which filters imitating the diaphragm/bell functions, and storage ability for the recorded sound signals [18, 19]. Thus, the digital electronic stethoscope amplifies the body sounds to strengthen the low sound levels with an economical recordable ability. The hardware design of the electronic stethoscope includes a power supply, acoustic sensor, preamplifier, low-pass filter, and power amplifier. In addition, an audio output can be included to display the phonocardiogram, if it will be used to hear the heart sounds.

4.4 Acoustic Wave Sensors

Piezoelectricity is a coupling between electric polarization and elastic deformation, which occurs in definite crystals, such as sapphire, lithium niobate, and quartz. Acoustic wave sensors are commonly designed using quartz crystal microbalance. The quartz crystal microbalance's fundamental principle uses piezoelectricity to convert mechanical and electrical signals. The surface acoustic wave (SAW) devices are enormously common in several electronics forms as band-pass filters. Entirely, the acoustic wave devices are sensors, which are sensitive to several physical parameter perturbations. A change in the output occurs with the change in the path characteristics through which the acoustic wave propagates. Several acoustic wave sensors generate propagating waves in shear-horizontal (SH) wave. The SH motion does not emit substantial energy into liquids, which prevents the damping during the liquid operation. However, the SAW sensor has an extensive displacement on the surface-normal direction, which releases compression waves producing extreme damping. This rule is applied in all cases except in the case of the devices that use propagating waves at a velocity lesser than the velocity of sound in the liquid. Generally, there are different acoustic waves that can be used for sensors, including Lamb wave, surface-skimming bulk wave, and the flexural plate wave [20–22].

One of the important factors during the selection of the suitable sensor is the sensor's sensitivity, which is relative to the amount of energy in the disturbed path of propagation. From the surface, the bulk acoustic wave (BAW) sensors scatter the energy via the bulk material to the further surface. This energy distribution minimizes the energy density on the surface at which the sensing occurs. Conversely, on the surface, the SAW sensors concentrate their energy leading to sensitive surfaces. Several phenomena can be detected using acoustic wave devices after coating these devices with materials of changed conductivity, elasticity, and mass. Under an applied stress, these sensors can be force, torque, pressure, and shock detectors. When the particles contacted the propagation medium, they become gravimetric or mass sensors. The detector can be a biosensor if the coating absorbs definite organic chemicals in the liquids. Furthermore, by choosing the precise orientation of propagation, a wireless temperature sensor can be formed, where the output is affected by the temperature that changes the propagating medium [23].

4.4.1 Bulk Acoustic Wave Sensors

A bulk wave is defined as the propagating wave via the substrate. The shear-horizontal acoustic plate mode (SH-APM) sensor and thickness shear mode (TSM) resonator are the most used BAW devices. The TSM is known as a quartz crystal microbalance (QCM), which consists of AT-cut quartz thin disk with parallel circular electrodes on both sides. The crystal shear deformation occurs by applying a voltage between the electrodes. Since the crystal resonates as electromechanical, standing waves are produced; thus, this device is considered a resonator. At the crystal faces, the displacement is maximized, which makes the device sensitive to surface interactions. The TSM resonator is considered a biosensor, which is able to detect and measure the liquids due to its shear wave propagation component that operates in the range of 5 and 30 MHz. At higher frequencies, designing a very thin device raises the mass sensitivity. However, thinning the sensors further than its normal range leads to fragile devices, which are challenging fabrication issue [24–27].

The SH-APM device uses a thin plate/piezoelectric substrate acting as an acoustic waveguide to limit the energy between the plate's lower and upper surfaces. The detection can happen to both surfaces which undergo displacement, where one side encloses the interdigital transducers, which are isolated from the conducting gases/fluids, whereas the other side is used as the sensor.

The BAW biosensors engage either shear or longitudinal waves. However, in the medium of interest, the shear-based acoustic wave biosensors are desired as they reduce the acoustic radiation. The BAW devices entail parallel electrodes located on the both sides of the crystal's thin piece. The BAW sensors can precisely use any piezoelectric element, especially the quartz due to its inexpensive costs and its simple synthesizes. Moreover, at high temperatures, the quartz thin disks are more stable compared to other piezoelectric materials. Between the biosensor two electrodes, potential differences and the crystal shear deformation occur, when applying an alternating electric field.

Consequently, across the quartz bulk, a mechanical standing wave oscillation occurs with a vibration frequency that depends on the quartz properties, including the phase, size, and density. Recently, the SH-APM sensor, the TSM resonator, the flexural plate wave sensors, and the thin rod acoustic wave sensors are the most prevalent BAW sensors [28].

4.4.2   Surface Acoustic Wave Sensors

The SAW travels near/along the piezoelectric material surface dissimilar to the BAW, which interacts only with the environment at the material opposite surface by traversing via it. The SAW sensors are used to measure the viscosity, temperature, chemical/biological entities, acceleration, pressure, and concentration. This acoustic wave sensor consists of piezoelectric substrate, electrodes (micrometallization patterns), active thin films, and interdigital transducers (IDT). The piezoelectric device senses small changes in the mass at the sensor surface, where the electrodes on the piezoelectric substrate surface transmit/receive acoustic waves. These waves are limited to the piezoelectric substrate surface, and the excited wave propagates along the crystal surface. Thus, the velocity of the surface wave changes due to the viscosity or mass changes. The SAW device's operating range depends on the crystal substrate acoustic velocity and IDT wavelength, which is ranging from MHz to the GHz.

A surface wave is the propagating wave on the substrate surface. The surface transverse wave (STW) sensor and the surface acoustic wave sensor are the broadly used surface wave devices. The Rayleigh waves have vertical and longitudinal shear components, which can be coupled with the contacting medium to the device surface. This coupling affects the velocity and amplitude of the wave, which empowers directly the SAW sensors to sense the mechanical and mass properties. Additionally, the SAW device can be used as microactuators due to its surface motion. The velocity of the generated wave has less magnitude compared to the consistent electro-

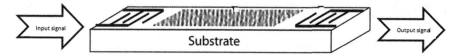

Fig. 4.1 Layout of the SAW sensor

magnetic wave by about five times. This makes the Rayleigh surface wave propagates in solid, where the wave amplitudes are about 10 Å with wavelength within the 1–100 microns range. The SAW sensor operates at frequencies from 25 to 500 MHz. However, its disadvantage is due to the Rayleigh waves, which are surface-normal waves that make the SAW sensors poorly suitable for liquid sensing, when contacted by a liquid as excessive surface wave attenuation occurs [29–32].

The technology of the SAW filter design is predominately fitted the infrastructure industry. Figure 4.1 illustrates the simple structure of the SAW device. It entails a IDTs pair of input/output transduction, and a delay path in the middle, which is coated with a sensing film to propagate the acoustic wave.

The SAW devices are used individually or in arrays as sensors in several applications including gas, vapor, and biological systems. Several acoustic wave devices are explicit to their operational phase. The SAW sensor's operating frequency is from 25 to 500 MHz. The main disadvantage of such devices is their unsuitability for liquid sensing due to the Rayleigh waves, which are surface-normal waves. Consequently, once the SAW sensor is contacted with liquid, an extreme surface wave attenuation occurs [33–35].

4.5 Acoustic Wave Propagation Modes

Commercially, acoustic wave filters/devices are widely used in medical applications, base stations, and mobile cell phones. In the transceiver electronics, the SAW devices can be considered as band-pass filters in both the intermediate frequency and radio frequency ranges. Sensors are one of the evolving applications of the SAW, including medical applications using chemical sensors, automotive applications using pressure sensors, and commercial/industrial applications using mass/temperature sensors. Acoustic wave sensors are very sensitive, intrinsically rugged, and essentially reliable. In addition, some of them can be wirelessly interrogated without any required sensor power.

The wave propagation mode is considered the descriptive characteristic of the acoustic wave devices. These waves are the ones that path over or on the piezoelectric material. Mainly, the acoustic waves are recognized by their displacement directions and velocities. Several arrangements are potential based on the boundary conditions and used material [36–40]. The integrated device technology (IDT) of each sensor offers the necessary electric field to relocate the material (substrate) for forming the acoustic wave. At the IDT on the other side, the wave transformed back into an electric field once it propagates through the substrate.

References

1. Corso, C. D. (2008). *Theoretical and experimental development of a ZnO-based laterally excited thickness shear mode acoustic wave immunosensor for cancer biomarker detection* (Doctoral dissertation, Georgia Institute of Technology).
2. Rocha-Gaso, M. I., March-Iborra, C., Montoya-Baides, Á., & Arnau-Vives, A. (2009). Surface generated acoustic wave biosensors for the detection of pathogens: A review. *Sensors, 9*(7), 5740–5769.
3. Wu, H. (2017). *Acoustic wave biosensors for biomechanical and biological characterization of cells* (Doctoral dissertation, University of Pittsburgh).
4. Campos, A. P. (2015). *CMOS integration of AlN based piezoelectric microcantilevers* (Doctoral dissertation, Universidad Politécnica de Madrid).
5. Lines, M. E., & Glass, A. M. (1977). *Principles and applications of ferroelectrics and related materials*. Oxford: Oxford University Press.
6. Harsányi, G. (2000). *Sensors in biomedical applications: Fundamentals, technology and applications*. Boca Rato: CRC Press.
7. Shen, B., Yang, X., & Li, Z. (2006). A cement-based piezoelectric sensor for civil engineering structure. *Materials and Structures, 39*(1), 37–42.
8. Wakekar, R. K., Bansod, N. N., Machindra, A. M., & Patil, S. R. (2015). Stampede alert system with heart beat sensor. *People, 24*.
9. Tian-Ling, R., Lin-Tao, Z., Li-Tian, L., & Zhi-Jian, L. (2002). Design optimization of beam-like ferroelectrics-silicon microphone and microspeaker. *IEEE Transactions on Ultrasonics, Ferroelectrics, and Frequency Control, 49*(2), 266–270.
10. Yoshino, Y., Makino, T., Katayama, Y., & Hata, T. (2000). Optimization of zinc oxide thin film for surface acoustic wave filters by radio frequency sputtering. *Vacuum, 59*(2–3), 538–545.
11. Saleh, S., Elsimary, H., Zaki, A., & Ahmad, S. (2006). Design and fabrication of piezoelectric acoustic sensor. *WSEAS Transactions on Electronics, 3*(4), 192.
12. De Silva, C. W. (2015). *Sensors and actuators: Engineering system instrumentation*. Boca Raton: CRC Press.
13. Lynch, J. P., & Loh, K. J. (2006). A summary review of wireless sensors and sensor networks for structural health monitoring. *Shock and Vibration Digest, 38*(2), 91–130.
14. Livingston, F. J. (2014). *Technology for improving the quality of life for patients suffering from vascular insufficiency*. North Carolina State University.
15. Newnham, R. E., Skinner, D. P., & Cross, L. E. (1978). Connectivity and piezoelectric-pyroelectric composites. *Materials Research Bulletin, 13*(5), 525–536.
16. Tressler, J. F., Alkoy, S., & Newnham, R. E. (1998). Piezoelectric sensors and sensor materials. *Journal of Electroceramics, 2*(4), 257–272.
17. Safari, A. (1994). Development of piezoelectric composites for transducers. *Journal de Physique III, 4*(7), 1129–1149.
18. Dolan, T. G., Oliver, S. R., & Maurer, J. F. (2001). Stethoscopes: Real-ear measurements and digital frequency transposition. *The Hearing Journal, 54*(1), 36–44.
19. Yashaswini, B. S., & Satyanarayana, B. S. (2012). The design of an electronic stethoscope-review. In *International conference on computer science and informatics (ICCSI)–Hyderabad–on 9th march*.
20. Grate, J. W., & Frye, G. C. (1996). Acoustic wave sensors. *Sensors Update, 2*(1), 37–83.
21. Ballantine, D. S., Jr., White, R. M., Martin, S. J., Ricco, A. J., Zellers, E. T., Frye, G. C., & Wohltjen, H. (1996). *Acoustic wave sensors: Theory, design and physico-chemical applications*. Elsevier.
22. Thompson, M., Kipling, A. L., Duncan-Hewitt, W. C., Rajaković, L. V., & Čavić-Vlasak, B. A. (1991). Thickness-shear-mode acoustic wave sensors in the liquid phase. A review. *Analyst, 116*(9), 881–890.
23. Vellekoop, M. J. (1998). Acoustic wave sensors and their technology. *Ultrasonics, 36*(1–5), 7–14.

24. Wohltjen, H. (1984). Mechanism of operation and design considerations for surface acoustic wave device vapour sensors. *Sensors and Actuators, 5*(4), 307–325.

25. Thompson, M., Dhaliwal, G. K., Arthur, C. L., & Calabrese, G. S. (1987). The potential of the bulk acoustic wave device as a liquid-phase immunosensor. *IEEE Transactions on Ultrasonics, Ferroelectrics, and Frequency Control, 34*(2), 127–135.

26. Peng, H., Liang, C., Zhou, A., Zhang, Y., Xie, Q., & Yao, S. (2000). Development of a new atropine sulfate bulk acoustic wave sensor based on a molecularly imprinted electrosynthesized copolymer of aniline with o-phenylenediamine. *Analytica Chimica Acta, 423*(2), 221–228.

27. Rey-Mermet, S., Lanz, R., & Muralt, P. (2006). Bulk acoustic wave resonator operating at 8 GHz for gravimetric sensing of organic films. *Sensors and Actuators B: Chemical, 114*(2), 681–686.

28. Lucklum, R., & Hauptmann, P. (2006). Acoustic microsensors—The challenge behind microgravimetry. *Analytical and Bioanalytical Chemistry, 384*(3), 667–682.

29. Shiokawa, S., & Kondoh, J. (2004). Surface acoustic wave sensors. *Japanese Journal of Applied Physics, 43*(5S), 2799.

30. Gronewold, T. M. (2007). Surface acoustic wave sensors in the bioanalytical field: Recent trends and challenges. *Analytica Chimica Acta, 603*(2), 119–128.

31. Wohltjen, H. (1984). Mechanism of operation and design considerations for surface acoustic wave device vapour sensors. *Sensors and Actuators, 5*(4), 307–325.

32. Josse, F., Bender, F., & Cernosek, R. W. (2001). Guided shear horizontal surface acoustic wave sensors for chemical and biochemical detection in liquids. *Analytical Chemistry, 73*(24), 5937–5944.

33. Länge, K., Rapp, B. E., & Rapp, M. (2008). Surface acoustic wave biosensors: A review. *Analytical and Bioanalytical Chemistry, 391*(5), 1509–1519.

34. Grate, J. W. (2000). Acoustic wave microsensor arrays for vapor sensing. *Chemical Reviews, 100*(7), 2627–2648.

35. Hierlemann, A., Zellers, E. T., & Ricco, A. J. (2001). Use of linear solvation energy relationships for modeling responses from polymer-coated acoustic-wave vapor sensors. *Analytical Chemistry, 73*(14), 3458–3466.

36. Barkan, A., Merlino, R. L., & D'angelo, N. (1995). Laboratory observation of the dust-acoustic wave mode. *Physics of Plasmas, 2*(10), 3563–3565.

37. Mortet, V., Elmazria, O., Nesladek, M., Assouar, M. B., Vanhoyland, G., D'Haen, J., D'Olieslaeger, M., & Alnot, P. (2002). Surface acoustic wave propagation in aluminum nitride-unpolished freestanding diamond structures. *Applied Physics Letters, 81*(9), 1720–1722.

38. Ohmachi, Y., Uchida, N., & Niizeki, N. (1972). Acoustic wave propagation in TeO 2 single crystal. *The Journal of the Acoustical Society of America, 51*(1B), 164–168.

39. Wohltjen, H., & Dessy, R. (1979). Surface acoustic wave probe for chemical analysis. I. Introduction and instrument description. *Analytical Chemistry, 51*(9), 1458–1464.

40. Engan, H. E., Kim, B. Y., Blake, J. N., & Shaw, H. J. (1988). Propagation and optical interaction of guided acoustic waves in two-mode optical fibers. *Journal of Lightwave Technology, 6*(3), 428–436.

Chapter 5
Acoustic Sensors in Biomedical Applications

Abstract The biomedical engineering domain is concerned with physiological modeling, biomaterials, biomechanics, control and simulation, etc. Biomedical sensors are considered the most vital parts in the biomedical engineering. These sensors enable the biologic events detection and conversion to signals. The biomedical sensors receipt signals that represent the biomedical measurements and convert them into optical or electrical signals. Thus, the biomedical sensor acts as an interface between the biological feature and the electronic system. Sensor specialists and biomedical engineers are interested to process and design sensors for several application problems. This chapter introduces some examples of the acoustic sensors in different biomedical applications.

5.1 Acoustic Waveguide Sensor for Chemical Detection

Biosensors can detect chemicals in the body liquids, which are fabricated using the thickness shear mode (TSM) resonator, which is a BAW device, and the SH-SAW sensors. This is due to the fact that the SAW device has attenuated waves and poor performance in liquids as the propagating wave's vertical component will be blocked by the liquid. Additionally, for liquid sensing, the Love wave acoustic sensor, which is the superior class of the SH-SAW, has the maximum sensitivity [1]. In addition, a complete biosensor is formed by placing a bio-recognition coating on the waveguide coating. Gizeli et al. [2] designed a direct immuno-sensor using a transducer based on acoustic wave device of acoustic waveguide geometry supporting the Love wave. A gold layer was used, where the bio-recognition surface formed on. The adapted surface was employed as an immuno-sensor model, which effectively sensed rabbit anti-goat IgG in the $3 \times 10{-}8{-}10{-}6$ M concentration range. Each binding step's specificity has been considered by the acoustic wave device.

In the Love wave device, on a substrate, the shear wave transmits in the low shear acoustic velocity material (upper layer) with a higher shear acoustic velocity, thus,

N. Dey et al., *Acoustic Sensors for Biomedical Applications*,
SpringerBriefs in Electrical and Computer Engineering,
https://doi.org/10.1007/978-3-319-92225-6_5

representing bilayer geometry. At a specific frequency, the Love wave device provides huge design suppleness, where the energy limitation is determined by deposited overlayer thickness and the acoustic properties. The inelastic polymer layer on top of the quartz substrate guarantees the sensitivity of the waveguide surface structure. This device can work in the liquid existence without losses owing to the mode conversion of the wave shear nature. In the liquid sample, within about 60 nm from the device surface, the evanescent field of the shear acoustic wave probes' electric, viscosity, and mass changes occurs. Consequently, by monitoring the acoustic wave propagation characteristics, including the frequency, phase, and amplitude, it is probable to detect the binding kinetics and obtain the corresponding acoustic to the optical immuno-sensor.

5.2   Stethoscopic Sensor for Respiratory Sound Recording

On the chest surface, the respiratory sounds can be recorded for diagnosis within the frequency ranges from 70–80 to 1000 Hz. For recording the respiratory sounds, there are three acoustic sensor types that can be in contact with the body surface, namely, the so-called contact sensors [3], stethoscopic sensors with microphones [4], and the acoustic accelerometers [5, 6], where a sensitive piezo-element is placed between the chest's surface and the housing. Adequate accurate mechanic-acoustical undistorted conversion of the phase-frequency and amplitude-frequency characteristics of the used acoustic sensors is essential. Such required characteristics in the 100–700 Hz frequency range can be achieved by employing an oscillatory displacement receivers founded on heavy stethoscopic sensors. This arrangement is supported with a microphone to be fixed in the neck along with a dynamic force receivers founded on heavy sensors with a longitudinal deformed piezo-transducer between the sensor housing and the body surface as proposed by Korenbaum et al. [3]. The authors analyzed a theoretical model, including a receiver and a vibrational system representing the biological tissues with concentrated parameters. A comparison between the model estimations and the results of the sensor's characteristics on the surface of the chest was conducted. There are different components of the proposed system for the respiratory sounds recording using the stethoscopic sensor [3].

Korenbaum et al. designed a stethoscopic sensor of the following parts: (a) the stethoscopic sensor device, including a microphone with preamplifier and stethoscopic attachment; (b) the light accelerometer device, including the bimorphic piezo-element, the base, the cable input, the preamplifier, and the housing; (c) the contact sensor device with longitudinal piezo-element, including the ring bearing element, the ring piezo-transducer, the heavy housing, and the housing lid; and (d) the contact sensor device with bent piezo-element, including the elastic plate, the round piezoplate, the housing, the cable input, and the preamplifier [3].

The stethoscopic sensor operation according to its parts can be explained as follows: a medical stethoscope is working with a built in microphone at frequencies of $f >> f_0$. A longitudinal wave travels from the chest due to the wave dimensions compactness of the chest-piece-ringed edge on the body surface as well as the biological tissues' small viscosity. The light accelerometer includes a bimorphic flexural piezo-element as a sensitive element, where this acoustic sensor under the resonance condition of its sensitive element during the operation at $f < f_0/(1.5 - 2)$ will produce accelerating vibrations with the chest surface. The contact acoustic sensor is displayed, where a sensitive piezo-element is placed between the chest surface and the sensor housing. Thus, sensors are created with a longitudinally deformed and flexural piezo-element, which represents a light sensor at $f >> f_0$ [3].

Generally, a common characteristic of all acoustic receivers' types is the determined deferment resonance, which depends on the sensor's mass and the biological tissues' hardness at the contact with the chest surface. Furthermore, the physical accuracy of the used acoustic sensors is owing to the following: (i) at frequencies higher than the suspension resonance, the stethoscopic receiver with a microphone acts as an oscillatory displacement receiver; (ii) at frequencies lower than the suspension resonance, the accelerometer acts as an oscillatory acceleration receiver; and (iii) at frequencies lower than the piezo-element's natural resonance and higher than the suspension resonance, the contact sensor acts as a dynamic force receiver.

5.3 Heart Sound Analysis in Clinical Diagnosis

The acoustic stethoscope is the main device to hear the HS. For diagnosing the cardiac disease, the direct auscultation is considered a conventional method. However, the HS interpretation requires sufficient physiologic knowledge about the cardiovascular system. The HS recording as a waveform form is called phonocardiogram (PCG) for visual inspect of the heart sounds. Valvular cardiac dysfunctions detection can be done using auscultation based on advanced signal processing techniques for acquisition to collect the HS samples and analysis of the HSs to diagnose the heart pathologic conditions [7].

Clinical diagnosis requires effective computer modeling of the diagnosis based on digital signal processing procedures to measure the data and its samples for further analysis as an overall system that shows what happens after acquiring such signals. The cardiovascular processes complexity is limitless, which requires mathematical models. Essentially, these processes rely on several transformations to characterize the system. The mechanical body processes create sounds that indicate the individual's health status to diagnose the patients with cardiovascular conditions. Any noteworthy variation in the PCG waveform is considered a sign or symptom of pathology [8].

During the heart valve closure, the normal heart sounds are produced, while the murmurs are produced by turbulent blood flow due to the leaking/narrowed valves, which is a public abnormal heart phenomenon. The murmurs are recognized from the basic HSs as they have a longer duration. Thus, the interpretation of the PCG waveform is essential to identify the heart abnormality cases. The phonocardiogram interpretation is a very challenging task owing to the parameters, which influence the HS generation and transmission along with the heart pathological condition that may not be identified in the raw time domain PCG signal, where for clinical diagnosis, the limitations of the PCG include [9] (i) the presence of noise and artifacts that mask the weak HSs, (ii) failed presentation of the HS information in the frequency components, (iii) absence of the energy variation information in the different sounds and the incapability to distinguish between the separated frequencies of different sounds, and (iv) complex identification of specific HS boundaries.

However, the PCG signals comprise features, such as the S1 representing the heart sound and S2 representing the location, the number of the sound components, and their time interval and frequency contents. These features can be measured using digital signal processing procedures. The HS analysis includes three phases, namely, (i) segmentation to identify the complete heartbeat borders, (ii) feature extraction to compute the distinctive parameters/characteristics of the cardiac cycle, and (iii) classification to determine the HS nature based on the distinct characteristics. Since the frequency spectrum characterizes the signal frequency components, the Fourier transform is used to determine the frequency-amplitude representation of a signal. Several studies have been conducted for clinical diagnosis of the heart sound signal analysis based on the Fourier transform [10]. Nevertheless, there is an overall consensus, lack of the research studies, and the inter-patient flexibility of signal processing procedures. In addition, the clinical validation of the analysis methods is insufficient, where measurement processes and data integrity are still in doubt. Consequently, several machine learning and soft computing techniques are raised to solve the different medical signal processing stages.

References

1. Kovacs, G., & Venema, A. (1992). Theoretical comparison of sensitivities of acoustic shear wave modes for (bio) chemical sensing in liquids. *Applied Physics Letters, 61*(6), 639–641.
2. Gizeli, E., Liley, M., Lowe, C. R., & Vogel, H. (1997). Antibody binding to a functionalized supported lipid layer: A direct acoustic immunosensor. *Analytical Chemistry, 69*(23), 4808–4813.
3. Korenbaum, V. I., Tagil'tsev, A. A., D'yachenko, A. I., & Kostiv, A. E. (2013). Comparison of the characteristics of different types of acoustic sensors when recording respiratory noises on the surface of the human chest. *Acoustical Physics, 59*(4), 474–481.
4. Korenbaum, V. I., Tagil'tsev, A. A., Kostiv, A. E., Gorovoy, S. V., & Pochekutova, I. A. (2008). Acoustic equipment for studying human respiratory sounds. *Instruments and Experimental Techniques, 51*(2), 296–303.
5. Korenbaum, V. I., Nuzhdenko, A. V., Tagiltsev, A. A., & Kostiv, A. E. (2010). Investigation into transmission of complex sound signals in the human respiratory system. *Acoustical Physics, 56*(4), 568–575.

6. Korenbaum, V. I., D'yachenko, A. I., Nuzhdenko, A. V., Lopatkin, N. S., Tagil'tsev, A. A., & Kostiv, A. E. (2011). Transmission of complex sound signals in the human respiratory system as a function of sound velocity in the utilized gas mixture. *Acoustical Physics, 57*(6), 872–879.
7. Emmanuel, B. S. (2012). A review of signal processing techniques for heart sound analysis in clinical diagnosis. *Journal of Medical Engineering & Technology, 36*(6), 303–307.
8. Martinez-Alajarin, J., & Ruiz-Merino, R. (2005, June). Efficient method for events detection in phonocardiographic signals. In *Bioengineered and bioinspired systems II* (Vol. 5839, pp. 398–410). International Society for Optics and Photonics.
9. Singh, J., & Anand, R. S. (2007). Computer aided analysis of phonocardiogram. *Journal of Medical Engineering & Technology, 31*(5), 319–323.
10. Clifford, G. D. (2002). *Signal processing methods for heart rate variability* (Doctoral dissertation, University of Oxford).

Chapter 6
Conclusion

This book introduces the basic definitions of the sensors, the biosensors and their features, and the equivalent components, amplifiers, filters, and bio-measurement systems for further circuit design. It describes and categorizes the mainstream acoustic wave biosensors, including the utilization of the bulk acoustic waves and analysis devices, which imply surface acoustic waves. In addition, the use of the piezoelectric substrates of the acoustic sensors design is included. The different types of the biosensors are presented. Several applications of the acoustic biosensors are introduced.

Biomedical signals represent physiological activities and observations of the different living organisms extending from protein/gene sequences to cardiac rhythms, tissues, and organs. Biosignal is any signal that transduced from a medical or biological source ranging from the cell level, molecular level, or organic/systemic level. In the clinics and laboratories, a broad diversity of such signals is usually encountered. Such biosignals include the speech signals, the ECG, the EEG, the electroneurogram, and the EMG. These biomedical signals are clinically acquired to monitor (detect/estimate) a specific physiological/pathological conditions for diagnosis and therapy. Furthermore, acquisition of multiple biosignal channels is available leading to extra challenges in the biosignal processing methods to measure the physiological meaning of the interactions between the different channels.

The processing of these biosignals aims to extract the imperative information from the biosignals after filtering to remove noise using biomedical signal processing algorithms. Since the organisms are multifaceted systems as their subsystems interact together, the measured biosignals typically hold the other subsystems' signals. Thus, removing the unwanted signal components is essential. The ultimate techniques for noise cancelation analyze the signal spectra to suppress the unwanted frequency components. Generally, the noise that affects the biosignals arises from the biological systems under study, the electronic instruments, and the power line interference.

N. Dey et al., *Acoustic Sensors for Biomedical Applications*,
SpringerBriefs in Electrical and Computer Engineering,
https://doi.org/10.1007/978-3-319-92225-6_6

For remote monitoring, sensors without operating power requirements are extremely desirable while measuring and sensing temperature, for example. Typically, sensors have different applications including measuring viscosity, angular rate, shock, acceleration, force, blood flow, and the displacement. Additionally, the sensors have an acoustoelectric sensitivity to detect the ionic contaminants, the electric fields, and the pH levels. In healthy and diseased situations, the only information source to describe the human body functionality is the biosensors to detect and convert the physiological measurements to electrical ones. The obtained biosignals are natural and original. Researchers are interested to develop new biosensors for processing, measurement, interpretation, and analysis. Typically, a biosensor can be defined as any hardware component which interacts with the physiological or biological system to obtain a medical signal for diagnosis and therapy. After gathering the medical data using the biosensors, processing phases are applied using biosignal processing methods toward automated interpretation. Since the body sends out weak electrical signals that are captured and transformed into information, researchers are interested to isolate the noisy signal that affected by other body signals to deliver a real-time display of the biosignal under concern.

Acoustic wave sensors are enormously adaptable devices, which are very sensitive, intrinsically reliable, inherently rough, and can be interrogated wirelessly and passively. Generally, the SAW sensors have verified to be the most sensitive due to their large energy density on the surface. However, for liquid sensing, the SH-APM sensors called the Love wave sensors demonstrated to be the most sensitive. Recently, several studies and extensive research work are continuing to develop these sensors for future medical applications and to design sensors with less power consumption with long battery lifetime and to handle the high attenuation caused by the body tissues.

There are numerous applications of the sensors in all domains. The SAW technology was used initially in the pressure sensor fabrication, where their velocities are affected strongly by applying stress to the piezoelectric substrate through which the wave is propagating. These SAW pressure sensors are rugged, extremely small, wireless, and do not require power (passive). In addition, the temperature sensor can be fabricated from crystalline material, where the surface wave velocities depend on the temperature and can be determined according to the used material type and orientation. Based on the SAW delay line oscillators, this sensor can have low hysteresis, high resolution of millidegrees, and good linearity. Nevertheless, this temperature sensor is very sensitive to the mass loading. Recently, the, surface-skimming, ST-cut quartz bulk wave proved its efficiency as a temperature sensor with less sensitivity to the mass loading compared to the SAW sensors.

In the medical domain and healthcare, the acoustic sensors have a vital role. The acoustic monitoring technology has several applied applications in the wearable devices and to monitor the patients acoustically [1–3]. Furthermore, the wireless acoustic sensors have been used in the phonocardiograph, which is an instrument for recording the heart sounds of the pumping action. Wireless phonocardiography sensors in the heart sound acquisition become essential [4–6]. The acoustic sensors have also another medical application for detecting the coronary artery diseases acoustically [7] and in the hearing aid application [8–11]. The machine learning techniques play an imperative role in the processing phases of the captured signals by the acoustic sensors in the different applications [12–17].

References

1. Gopalsamy, C., Park, S., Rajamanickam, R., & Jayaraman, S. (1999). The Wearable Motherboard™: The first generation of adaptive and responsive textile structures (ARTS) for medical applications. *Virtual Reality, 4*(3), 152–168.
2. Penner, A., Doron, E., & Porat, Y. (2001). *U.S. Patent No. 6,198,965*. Washington, DC: U.S. Patent and Trademark Office.
3. Mba, D., & Rao, R. B. (2006). Development of acoustic emission technology for condition monitoring and diagnosis of rotating machines; bearings, pumps, gearboxes, engines and rotating structures. *The Shock and Vibration Digest, 38*(1), 3–16.
4. Chourasia, V. S., & Tiwari, A. K. (2012). Wireless data acquisition system for fetal phonocardiographic signals using BluetoothTM. *International Journal of Computers in Healthcare, 1*(3), 240–253.
5. Sa-Ngasoongsong, A., Kunthong, J., Sarangan, V., Cai, X., & Bukkapatnam, S. T. (2012). A low-cost, portable, high-throughput wireless sensor system for phonocardiography applications. *Sensors, 12*(8), 10851–10870.
6. Zhang, Y., Fogoros, R., Haro, C., Dalal, Y., Brockway, M., & Siejko, K. Z. (2011). *U.S. Patent No. 7,922,669*. Washington, DC: U.S. Patent and Trademark Office.
7. Semmlow, J., & Rahalkar, K. (2007). Acoustic detection of coronary artery disease. *Annual Review of Biomedical Engineering, 9*, 449–469.
8. Miles, R. N., & Hoy, R. R. (2006). The development of a biologically-inspired directional microphone for hearing aids. *Audiology and Neurotology, 11*(2), 86–94.
9. Bertrand, A., & Moonen, M. Robust distributed noise reduction in hearing aids with external acoustic sensor nodes. *EURASIP Journal on Advances in Signal Processing, 2009, 2009*, 12.
10. Ko, W. H., Zhang, R., Huang, P., Guo, J., Ye, X., Young, D. J., & Megerian, C. A. (2009). Studies of MEMS acoustic sensors as implantable microphones for totally implantable hearing-aid systems. *IEEE Transactions on Biomedical Circuits and Systems, 3*(5), 277–285.
11. Doclo, S., Gannot, S., Moonen, M., & Spriet, A. (2010). Acoustic beamforming for hearing aid applications. In *Handbook on array processing and sensor networks* (pp. 269–302). Hoboken: Wiley-IEEE.
12. Maglogiannis, I., Loukis, E., Zafiropoulos, E., & Stasis, A. (2009). Support vectors machine-based identification of heart valve diseases using heart sounds. *Computer Methods and Programs in Biomedicine, 95*(1), 47–61.
13. Hu, T., & Fei, Y. (2010). QELAR: A machine-learning-based adaptive routing protocol for energy-efficient and lifetime-extended underwater sensor networks. *IEEE Transactions on Mobile Computing, 9*(6), 796–809.
14. Yatani, K., & Truong, K. N. (2012, September). BodyScope: A wearable acoustic sensor for activity recognition. In *Proceedings of the 2012 ACM Conference on Ubiquitous Computing* (pp. 341–350). ACM.
15. Palaniappan, R., Sundaraj, K., & Ahamed, N. U. (2013). Machine learning in lung sound analysis: A systematic review. *Biocybernetics and Biomedical Engineering, 33*(3), 129–135.
16. Özdemir, A. T., & Barshan, B. (2014). Detecting falls with wearable sensors using machine learning techniques. *Sensors, 14*(6), 10691–10708.
17. Lane, N. D., Georgiev, P., & Qendro, L. (2015, September). DeepEar: Robust smartphone audio sensing in unconstrained acoustic environments using deep learning. In *Proceedings of the 2015 ACM International Joint Conference on Pervasive and Ubiquitous Computing* (pp. 283–294). ACM.

Index

Printed in the United States
By Bookmasters